Data Analysis and Methods of Qualitative Research:

Emerging Research and Opportunities

Silas Memory Madondo
CeDRE International Africa, Zimbabwe

A volume in the Advances in Library and Information Science (ALIS) Book Series

IGI Global
PUBLISHER of TIMELY KNOWLEDGE

Published in the United States of America by
IGI Global
Information Science Reference (an imprint of IGI Global)
701 E. Chocolate Avenue
Hershey PA, USA 17033
Tel: 717-533-8845
Fax: 717-533-8661
E-mail: cust@igi-global.com
Web site: http://www.igi-global.com

Library of Congress Cataloging-in-Publication Data

Names: Madondo, Silas Memory, 1976- author.
Title: Data analysis and methods of qualitative research: emerging research and opportunities / by Silas Memory Madondo.
Description: Hershey, PA : Information Science Reference, [2021] | Includes bibliographical references and index. | Summary: "This authored book examines practical techniques for researchers undertaking qualitative projects that are not contaminated with quantitative or mixed methods research methods and concepts, by discussing in detail methods related to social constructions paradigms"-- Provided by publisher.
Identifiers: LCCN 2021005784 (print) | LCCN 2021005785 (ebook) | ISBN 9781799885498 (hardcover) | ISBN 9781799885504 (paperback) | ISBN 9781799885511 (ebook)
Subjects: LCSH: Qualitative research--Methodology. | Qualitative research--Data processing. | Social sciences--Resarch.
Classification: LCC H62 .M116 2021 (print) | LCC H62 (ebook) | DDC 001.4/2--dc23
LC record available at https://lccn.loc.gov/2021005784
LC ebook record available at https://lccn.loc.gov/2021005785

This book is published in the IGI Global book series Advances in Library and Information Science (ALIS) (ISSN: 2326-4136; eISSN: 2326-4144)

British Cataloguing in Publication Data
A Cataloguing in Publication record for this book is available from the British Library.

Advances in Library and Information Science (ALIS) Book Series

ISSN:2326-4136
EISSN:2326-4144

Editor-in-Chief: Alfonso Ippolito, Sapienza University-Rome, Italy; Carlo Inglese, Sapienza University-Rome, Italy

MISSION

The **Advances in Library and Information Science (ALIS) Book Series** is comprised of high quality, research-oriented publications on the continuing developments and trends affecting the public, school, and academic fields, as well as specialized libraries and librarians globally. These discussions on professional and organizational considerations in library and information resource development and management assist in showcasing the latest methodologies and tools in the field.

The **ALIS Book Series** aims to expand the body of library science literature by covering a wide range of topics affecting the profession and field at large. The series also seeks to provide readers with an essential resource for uncovering the latest research in library and information science management, development, and technologies.

COVERAGE

- Knowledge Management Learning (Course) Management Software
- Patents/Trademark Services
- Librarianship and Human Rights
- Information Literacy
- ©/Fair Use
- Outsourcing of Library Services
- Research Habits of Students and Faculty
- Gaming in Libraries
- Collaborative Learning and Libraries
- Building and Space Planning

The Advances in Library and Information Science (ALIS) Book Series (ISSN 2326-4136) is published by IGI Global, 701 E. Chocolate Avenue, Hershey, PA 17033-1240, USA, www.igi-global.com. This series is composed of titles available for purchase individually; each title is edited to be contextually exclusive from any other title within the series. For pricing and ordering information please visit http://www.igi-global.com/book-series/advances-library-information-science/73002. Postmaster: Send all address changes to above address.

Titles in this Series

For a list of additional titles in this series, please visit: http://www.igi-global.com/book-series

Handbook of Research on Records and Information Management Strategies for Enhanced Knowledge Coordination
Collence Takaingenhamo Chisita (Department of Information Science, University of South Africa, South Africa) Rexwhite Tega Enakrire (Department of Information Science, University of South Africa, South Africa) Oluwole Olumide Durodolu (Department of Information Science, University of South Africa, South Africa) Vusi Wonderboy Tsabedze (Department of Information Science, University of South Africa, South Africa) and Joseph M. Ngoaketsi (Department of Information Science, University of South Africa, South Africa)
Information Science Reference • © 2021 • 538pp • H/C (ISBN: 9781799866183) • US $295.00

Big Data Applications for Improving Library Services
Sangeeta Namdev Dhamdhere (Modern College of Arts, Science, and Commerce, Pune, India)
Information Science Reference • © 2021 • 211pp • H/C (ISBN: 9781799830498) • US $195.00

Transforming Library Operations With ICT Tools
Okon Edet Ani (University of Calabar, Nigeria)
Information Science Reference • © 2021 • 300pp • H/C (ISBN: 9781799842798) • US $195.00

Open Access Implications for Sustainable Social, Political, and Economic Development
Priti Jain (University of Botswana, Botswana) Nathan Mnjama (University of Botswana, Botswana) and O. Oladokun (University of Botswana, Botswana)
Information Science Reference • © 2021 • 360pp • H/C (ISBN: 9781799850182) • US $195.00

IGI Global
PUBLISHER of TIMELY KNOWLEDGE

701 East Chocolate Avenue, Hershey, PA 17033, USA
Tel: 717-533-8845 x100 • Fax: 717-533-8661
E-Mail: cust@igi-global.com • www.igi-global.com

To my wife, Thelma, and our children, Ethan and Elsie.

Table of Contents

Preface

Dear Reader

- Are you an undergraduate/graduate student, a researcher or working for a research organization?
- Do you have problems in undertaking ideal qualitative projects that are free from quantitative methods and concepts?
- Do you have problems in applying ethics in qualitative projects?
- Do you have enough knowledge about the social constructionist philosophy and its application to qualitative projects?
- Do you want to acquire practical skills of handling and applying social constructionist designs in qualitative projects?
- Do you know the data gathering methods and instruments for qualitative projects?
- Do you want to acquire practical skills of handling and applying non-probability sampling procedures in qualitative projects?
- Do you know the steps of qualitative data analysis?
- Do you understand the concepts; constant comparative, analytic induction and the genesis of coding in qualitative projects?
- Do you want to gain skills of transcribing and transcription of qualitative data?
- Do you want to perform open, axial and selective coding within and outside the grounded theory?
- Do you know the strategies of boosting trustworthiness of qualitative projects and their outcomes?
- Do you want to know the structure and the writing of qualitative reports?

If you are currently experiencing some or all of the challenges highlighted above, then this book is an appropriate prescription for you. Data Analysis & Methods of Qualitative Research consists of eight chapters. The general purpose of this book is to enable the reader to undertake ideal qualitative projects that are free from quantitative research techniques/concepts. The reader will be able to handle ethical issues in qualitative research and acquire practical skills of handling ethics in qualitative projects. The discussion of research philosophies is critical in research. The reader will be able to apply the social constructionist paradigm in qualitative research projects. The book discusses in detail methods related to social constructionist paradigm that are popular with qualitative projects. In a nutshell, the book covers the following chapters: "Ethics in Qualitative Research," Chapter 1; "Qualitative Research Philosophy," Chapter 2; "Qualitative Research Designs," Chapter 3; "Qualitative Data Gathering Instruments and Methods," Chapter 4; "Qualitative Sampling Methods," Chapter 5; "Trustworthiness of Qualitative Research Outputs," Chapter 6; "Introduction to Qualitative Data Analysis," Chapter 7; and "The Structure and Writing of Qualitative Reports, Chapter 8. By the end of reading this book, readers should be able to become competent qualitative research specialists.

The chapters of the book are divided into seven sections and the sections are chapter abstract, key concepts, learning outcomes, a case study, an introduction, content and concluded by a question and answer session. Important and critical questions are suggested and answered in the question and answer section at the end of each chapter. The questions and answers are crafted to help the readers to understand the contents of the chapters in depth. The book was written following the author's interactions with researchers, undergraduate, graduate research and statistics students for more than 13 years. The book was written with the aim of helping researchers to understand the handling of ideal qualitative research projects. The author noted that quite a number of undergraduate/graduate students and researchers are finding it hard to separate qualitative research projects from quantitative research projects.

Chapter 1, "Ethics in Qualitative Research," introduces readers to research ethics that should be considered at various phases of undertaking qualitative projects. The ethics related to the selection of research title/problem are discussed in detail. The researchers are cautioned to visit the research sites prior to the adoption of their project titles/problems. The discussion of intellectual property rights should be done at this stage for the projects undertaken by two or more authors. The discussion of intellectual property rights is done to avoid future conflicts that may be provoked by the financial rewards attached to the projects.

The readers are exposed to the strategies of handling non-humans in qualitative research projects dealing with animals. The researchers undertaking their projects with non-humans (animals) should handle them with care and according to the dictates of animal rights. Other ethics related to the selection of research title/problem discussed in chapter one are ethics involving the handling of children in qualitative projects and ethics dealing with sensitive issues like drugs, religions and corruption. The readers will be able to acquire skills of handling ethics related to the research title/problem selection.

Chapter 1 also discusses ethics related to data gathering/collection. The section contains information related to the requesting of permission (from the research sites) to undertake qualitative research projects. Another critical element discussed is the informed consent agreement, the researchers may need to sign informed consent agreements with research subjects before starting collecting data from the research sites. The informed consent agreement may be verbal if the research subjects are not able to read and write. There are also rules related to the photographing done within the research sites. The readers will be able to understand the rules of photographing in this section.

Ethics connected to data analysis, report writing and dissemination of qualitative projects' outputs are discussed in chapter one. The readers will get insights on plagiarism, confidentiality, duration of keeping and destroying data from the research sites. The chapter is concluded by an article written by the author. The articles equips readers with practical experiences of handling ethics in ethnographic studies. The readers should be able to have insights of how qualitative researchers are handling ethics in their projects.

Chapter 2 is "Qualitative Research Philosophy." Just like quantitative and mixed methods research paradigms, qualitative research has its own philosophy. This chapter introduces readers to the social constructionist philosophy, a philosophy claimed to be behind qualitative research. The chapter is divided into several section and readers should be able to understand the genesis of social constructionist philosophy. They should be able to identify reasons for the emergence of the social constructionist philosophy and the paradigm wars between the positivists and the social constructionists. There is a section on the traits of social constructionist philosophy based on ontology, epistemology, axiology, methodology and rhetoric. The characteristics of qualitative research projects and key components of idea qualitative projects are highlighted in this chapter. Chapter two is concluded by a question and answer session where important questions on research philosophy are suggested and answered.

The aim of Chapter 3, "Qualitative Research Designs," is to equip the readers with appropriate designs for qualitative projects. The section discusses BIG five qualitative traditions/designs and others. The readers should be able to understand qualitative designs, identify data collection methods/sampling procedures linked to each design, grasp methods of data analyses under each design and to gain knowledge of steps of applying these designs to practice. Biography design is discussed in the first section and information related to the types of biography, its origin, steps of handling it in qualitative projects and the techniques of analyzing data collected via biography design.

The second section of Chapter 3 contains information related to phenomenology design. The readers should be able to describe the purpose of phenomenology, understand the procedures of administering interviews for phenomenological projects and the strategies of analyzing the gathered data. The third section contains information related to grounded theory and the readers will be able to know why grounded theory is unique. The analysis of data for grounded theory projects is also unique and readers are expected to appreciate this position. Case study design, historical design and ethnographic design are discussed in sections four, five, and six respectively. Chapter 3 is concluded by a question and answer session, the aim of the question and answer session is to enable readers to handle critical questions connected to qualitative research designs.

Chapter 4 is "Qualitative Data Gathering Instruments and Methods." Qualitative data collection methods and instruments are different from quantitative ones. The chapter discusses four key data collection methods for qualitative projects and their instruments. The readers will be able to know the uniqueness of qualitative research data collection methods and instruments. The aim of this chapter is to enable researchers to design qualitative data collection tools with minimum challenges and will be able to link data collection methods and instruments to qualitative research designs and social constructionist philosophy.

The first section of Chapter 4 contains information related to interview method in qualitative projects. The types of interviews are critically examined and the processes of administering interviews are highlighted. The readers will be able to acquire skill of handling interviews and designing interview guide. The second section contains information related to group interview (focus group discussion method). The readers will be able to know the distinction between interview method and group interviews in qualitative research projects. The techniques of administering group interviews are also discussed in this section.

Observation and document data gathering methods in qualitative projects are discussed in sections three and four respectively. Participant and non-participant observation are discussed and readers will be able to know the contexts of applying participant and non-participant observation methods in qualitative research projects. Primary and secondary documents are distinguished in this section and readers will be able to know the circumstances in which they are used in qualitative projects. The chapter is summarized by a question and answer session and readers are expected to broaden their understanding of data gathering methods and instruments. Critical questions related to data gathering methods and instruments are suggested, answered and readers will be able to understand the core contents of the chapter easily.

Chapter 5, "Qualitative Sampling Methods," discusses the non-probability sampling procedures that are connected to qualitative projects. The readers are expected to understand the contexts appropriate for the use of each and every sampling technique. The steps of selecting samples from the population are highlighted and practical examples are given. The chapter also discusses the controversial sampling procedures that are very difficult to classify like systematic, cluster and stratified sampling procedures.

The first section of Chapter 5 contains information connected to purposive sampling and its variants/types. The section discusses seven variants of purposive sampling and the variants are as follows: maximum variation, homogeneous, total population, expert, critical case study, deviant case and typical case sampling. The readers should be able to describe them, follow steps of applying them to practice, understand the contexts appropriate for adopting them and the designs related to them.

Snowball sampling is discussed in detail in Chapter 5 as well. Snowball sampling is unique because it is the only sampling method for qualitative projects that can be used to identify hidden samples. The readers will be able to define the term hidden sample, to understand the practical steps of applying snowball sampling and apply the procedure in qualitative research projects. The context appropriate for adopting snowball sampling is connected to its uniqueness. The sampling procedures for quantitative projects are not able to identify and select hidden respondents, only snowball sampling and other non-probability sampling procedures are able to deal with hidden populations. There are key ethical considerations for snowball sampling that are discussed in this chapter and readers should be able to explore them.

Quota sampling technique is another non-probability sampling procedure discussed in Chapter 5. The readers should be able to differentiate quota sampling from cluster and stratified sampling procedures and will also be able to understand the similarities among the three sampling procedures. The section discusses two major types of quota sampling procedure and their application. The steps of selecting elements from the population via quota sampling procedure are highlighted and readers will be able to know the contexts appropriate for adopting quota sampling procedure.

Self-selective sampling procedure is examined in Chapter 5 as well. The challenges of adopting it are highlighted and researchers should be able to control such challenges because they are capable of negatively affecting the credibility of qualitative projects and their outputs. The last and final qualitative sampling procedure discussed in chapter five is convenience sampling and steps of applying the sampling procedure to practice are highlighted. The readers should be able to apply the sampling procedure to appropriate contexts with minimum challenges. The chapter is concluded by a question and answer section. Core questions related to sampling procedures are raised and answered and readers will be able to understand the concepts related to sampling procedures for qualitative projects easily.

Chapter 6, "Trustworthiness of Qualitative Research Outputs," addresses issues related to the credibility, transferability, dependability and conformability of qualitative projects. Trustworthiness is the equivalence of reliability and validity in quantitative projects. The readers will be exposed to the strategies of promoting trustworthiness in qualitative projects. The chapter is concluded by the question and answer session answering critical questions related to trustworthiness.

Chapter 7, "Introduction to Qualitative Data Analysis," discusses the processes of coding of qualitative data in depth. The first section defines qualitative data analysis concept in light of segmentation and reassembling of data. Another component handled in chapter seven is the examination of the origin of coding in qualitative projects in relation to the constant comparative and analytic induction concepts. The readers will be able to learn the stages of qualitative data analysis in this chapter. The qualitative data analysis stages examined are; recording of qualitative data, transcription of qualitative data and coding of qualitative data before proceeding to qualitative report writing. The chapter is concluded by the practical examples of performing open, axial and selective coding within and outside the grounded theory.

Chapter 8, "The Structure and Writing of Qualitative Reports," is the last chapter of this book covering details related to the structure of qualitative projects and report writing. The readers will be enabled to handle the introduction and literature review of qualitative projects as well as the methodology, data analysis and discussion and summary, conclusion and recommendations.

Acknowledgment

I would like to express my gratitude to the undergraduate and graduate students that I taught research methods, statistics, and quantitative techniques courses from 2007 to 2018. The experiences and interactions with them provoked me to write this textbook. The aim of the textbook is to assist university students and researchers to undertake ideal qualitative projects with minimum challenges.

Special thanks goes to Dr. Kanga, A. she helped me to become a Mixed Methods Research specialist. I appreciate her coaching during the period that I was studying my PhD program, she managed to change my philosophy about research. My family was a source of inspiration throughout the period of writing this textbook, may they be blessed too.

Chapter 1
Ethics in Qualitative Research

ABSTRACT

This chapter exposes readers to practical techniques of handling ethics in qualitative research projects. Researchers will be able to understand qualitative research ethics for human and non-human research projects. The intellectual property discussion is central to qualitative research projects; readers will be exposed to the steps of undertaking intellectual property rights discussion. The chapter is divided into five sections; readers will be able to experience the dangers of overlooking ethics when undertaking qualitative research projects, and the chapter suggests possible solutions to reduce and control risks that are connected to the violations of research ethics.

1 KEY CONCEPTS

- **Ethics** – a study of what is right and wrong when undertaking qualitative research projects.
- **Humans** – people involved in research projects as respondents.
- **Non-humans** - animals or natural elements involved in research projects.
- **Gatekeepers** – leaders or people in charge of research sites, who help researchers to access research sites.
- **Key Informants** – people within the research sites who can provide reliable information or data to the researchers.
- **Intellectual Property** – connected to the ownership of the qualitative studies/outputs and the discussion related to the execution strategies.

DOI: 10.4018/978-1-7998-8549-8.ch001

- **Ethnography** – these are qualitative studies that deals with ethnic groups/tribes or peculiar groups in society.
- **Masaai & Hadzabe** – tribes found in Northern parts of Tanzania, still practising traditional cultural practices.
- **Reactivity** – reactions/conflicts/injuries/deaths that may occur during the period of data gathering in qualitative projects.
- **Reciprocity** – a symbiotic relationship between the researchers and the research subjects on research sites. I.e. the motivational benefits of the qualitative projects to the research subjects.
- **Bracketing** – the ability of researchers to suspend their emotions and feelings (avoid biases) when writing qualitative reports.
- **Grounded Theory** – a unique design for qualitative approach designed to develop theories.
- **Coding** – the regrouping of qualitative data into categories.

Learning Outcomes

By the end of the chapter, readers should be able to;

- Acquire skills of handling ethics connected to all phases of undertaking qualitative projects from research problem to report writing and dissemination of results.
- Follow and put into practice steps of intellectual property discussion.
- Understand ethics related to projects dealing with non-humans (animals).
- Handle unexpected moments (friction) in qualitative data gathering.
- Be emotionally provoked by some sad experiences reported.
- Experience how ethnographers are applying research ethics in their projects.
- Gain practical skills of dealing with ethical issues on research sites.

Case Study

On 2nd of October 2016, three workers of Selian Agricultural Research Institute were killed by Mvumi Ward Villagers in Dodoma, Tanzania. The researchers were on mission of data gathering. It was reported by the media that the villagers were shocked to see strangers in their village. The three researchers did not have gatekeepers to protect them from harm and deaths.

The villagers attacked them, they thought that the researchers were strangers. It was unfortunate that innocent researchers lost their lives because they did not follow proper research ethics.

2 INTRODUCTION

Undertaking research studies without adhering to ethics may be dangerous to researchers and research subjects. There are several cases reported of injury and deaths caused by the abuse of research ethics. According to the case study above, three research workers of Selian Agriculture Research Institute were murdered while on mission of data gathering. The duty of a researcher is to make sure that research subjects are protected from harm. Undertaking qualitative studies with communities is sometimes a risk activity. Research ethics are central to qualitative studies. The social constructionists (qualitative research specialists) criticized the positivists (quantitative research specialists) for ignoring research ethics. The first step for the researchers to do prior to data collection is to identify possible risks and suggest ways of handling them. If the risks are difficult to handle, the researchers are requested to abstain from such research activities.

There are ethics for humans and non-humans researches. Humans in the context of this study refer to people involved in the study as participants. Non-humans are animals or nature that can be involved in research studies. There are quite a number of studies that are undertaken with animals and nature. Humans and non-humans in qualitative research projects are treated as equals. They are all supposed to be protected from harm throughout the study. This section is divided into the following subheadings;

- Ethics Related to Research Problem
- Ethics for Data Collection
- Ethics for Data Analysis
- Ethics for Report Writing & Dissemination of Research Outputs
- Practical Experiences of Handling Ethics in Ethnographic Projects

Ethics Related to the Research Problem

The process of undertaking a research project starts with the selection of the research title which is popularly known as research problem in research. According to Punch (2005), researchers should select research problems that are beneficial to the communities and people involved. The research problems should benefit the individuals that are to be involved in the study. The research problems that are not valuable to communities and individuals (to be part of the study) should be avoided.

Researchers are encouraged to carry out pilot studies soon after identifying their research problems. A pilot study is done before the beginning of a project. It is done in order for the researchers to familiarize themselves with the research sites and will enable them to interact with the environment. There is need for researchers to undertake a situational analysis on the research sites before the commencement of their projects. According to Creswell and Clarke (2007), a pilot study is important because it establishes trust and respect between the researchers and the research subjects. The pilot studies will also enable the researchers to write credible research proposals that are supported with details of what is happening on the research sites.

After the identification of the research problems, researchers are encourage to handle issues and acquire skills related to intellectual property, handling animals, dealing with children and sensitive issues in research projects. These areas should be handled with extra care because they may cause harm to research subjects or even to researchers. It is therefore mandatory for the researchers to be equipped with skills of handling these sensitive matters in research.

Intellectual Property Discussion

Intellectual property means the ownership of the research outputs. It is the ownership of ideas and knowledge about the research as well as money that may be involved. Intellectual property in research is unique and complex in the sense that ideas, knowledge and research outputs are intangible assets. Intellectual property is different from land or residential properties because it is an intangible property and very complex to manage.

The researchers undertaking studies as a group should sit down and discuss intellectual property rights before progressing on the next step. Discussion of property rights will help to reduce conflicts among the researchers in future. A fair discussion of intellectual property is always good for peace and future progress. It should be noted that in some cases, research project involve the cash management. The distribution of cash among project team members may result in problems/conflicts if intellectual property discussion was ignored at the beginning of a project.

According to Smith (2003), researchers should sit down and thoroughly discuss intellectual property related to their research projects before the beginning of their projects. There are some university professors who are always taking advantage of their students. They use them in their research projects without discussing intellectual property rights. Such a practice is against the rules and dictates of the research ethics. This section discusses the steps that researchers should follow when discussing the issues related to intellectual property.

The first step is to make every researcher understand the meaning of intellectual property and the rationale behind discussions related to it. There are instances when some researchers are ignorant about intellectual property rights. The research team should discuss together several issues related to intellectual property. The researchers should discuss as equals and should agree on all issues raised. More time for discussion may be allowed in case of disagreements during the discussions.

The second step is the discussion of tasks and roles. The research team should begin by identifying all tasks that are related to the project. There are several tasks involved i.e. proposal writing, data collection, report writing and others. Every team member of the research project is supposed to be given tasks and roles to play during the period of undertaking a research project. The roles are very important because they are central to the execution of the research project.

The third step is the discussion of ethics. The research team should understand the ethics involved. They should understand the ethics on confidentiality, privacy and the rules of game. They should participate in the discussion of ethics and share their views on how possible risks involved may be handled. The researchers and research subjects should be protected from harm and a team work approach is needed when discussing issues related to ethics.

The fourth step is the discussions of issues related to the sharing of resources. If the research's project team members are supposed to contribute cash towards a project, the amount to be contributed by each member should be known. If the project is sponsored, the details about the sponsor should be exposed to the project team. The project team should also discuss about the distribution of the financial resources emanating from the project outputs. The members of the project should know the percentage of money that they should get in case of any payments. This is done in order to avoid conflicts in future. There may be other team members who may be working as volunteers, everything should be documented.

The final step will be to sign an agreement. Project team members should be given enough time to go through the document before agreeing to the terms. All members of the project team have to sign the agreement. This is done in order to avoid legal conflicts in future. The project may then proceed to another level after the signing of an agreement by the project team members. These are some steps that the project team may follow when discussing the intellectual property. However, the discussion arrangement may vary in terms of content depending on the nature of projects.

Dealing with Children

Research projects in social sciences often involve children apart from adults. After the selection of a research problem, the research team should check on whether their project will involve children as participants. If children are involved, the team should discuss on how to handle them throughout the project. In most countries of the World, a child is someone below the majority age. Majority age on average is 18 years and those people below the age of 18 are classified as children. Children include school kids, street kids and youngsters in special homes taking care of the abandoned children.

The first step for researcher to consider when dealing with children in their project, is to ask for the permission from the responsible authorities. According to Madondo (2016a), the people or bodies taking care of children may allow or deny researchers to go ahead with their projects. If the children are under the custodian of the schools, the school authorities will be responsible for giving or denying permission. If the children are within the parameters of children's homes, those in charge of them should be consulted for permission. In case of street kids, the local authorities in charge of the cities or towns involved should be consulted for permission.

The permission on involving kids in research projects should be followed by signing consent agreement forms. The researchers should avoid verbal agreements because they are weak at law. It is advisable for the researchers to draft consent agreement forms and sign them together with the institutions in charge of children. The consent agreement forms will be able to be used as evidence in courts of law in case of misunderstandings and legal disputes.

In a nutshell, dealing with children is a complex task because of child rights that are stipulated by the law. The researchers should make sure that these rights are protected. The violations of child rights may have a negative impact on the credibility of the projects. The credibility of research outputs may be questioned if child rights are violated. Therefore, the researchers should handle the children with extra care by respecting their rights.

Critical Thinking Challenge

1. You have been asked to undertake a study with primary school children, briefly describe how you would handle ethical issues related to children?

Dealing with Sensitive Research Areas

There are several sensitive areas in research. The researchers are therefore requested to assess the sensitivity of their research problems. Madondo (2016a) suggested drugs, religious matters and corruption as part of sensitive areas in research studies. These areas are sensitive because they may result in injury or loss of lives. The researchers should assess the risks that may be involved and take appropriate actions in order to solve the problems that may arise. The researchers should get skills on how to handle such matters before they move to the next level.

Dealing with drug dealers and subjects involved in corruption may be complex. In most cases, researchers may be harmed during the process of undertaking the projects that are sensitive. The people involved in corruption and drugs may fight back and the researchers need to be very careful. Despite the fact that corrupt and drug dealers are on the wrong side of the law, their confidentiality and privacy should be protected. Just like any research subjects, they should be protected by the research ethics. The researchers should avoid becoming spies in the process. They should not play the role of an investigator. The computers used to capture the details of such research subjects should be protected by computer passwords. Personal details should be concealed according to the rules and all research subjects should be protected irrespective of their conditions and previous details.

Dealing with sensitive research areas related to the drugs can be a challenge if the researchers are not experienced in these matters. The researchers may consider working with organizations or individuals dealing with drugs in order for the task to be accomplished with minimum problems. Such organisations and individuals are accessible to data and research sites. They are capable of providing reliable and credible data about the projects related to drugs. Religious matters are also sensitive because they are close to the souls of individuals. The researchers should make sure that their projects do not offend communities attached to certain religious perspectives. They should assess the risks that may be involved and handle them before moving to another level of the project.

Dealing with Non-Humans Animals

Animals and human beings are equal in research projects. *'There is no difference between humans and animals in qualitative projects. The only problem is that, humans think that they are the only creatures to inherit the kingdom of heaven. They will be shocked to see animals on the right side of God during the last hour'.* The rights for all (humans and non-humans) have to be protected throughout the period of undertaking research projects. The researchers should check on whether they are dealing with humans or non-humans. If they are dealing with animals, they should acquire some skills of handling animals before the undertaking of their projects.

According to Andre and Velasquez (2015) researchers planning to deal with animals in their projects should understand some fundamentals of dealing with animals. It is advisable for them to read literature of projects done with animals. They should work close to animal experts in order for them to be successful. This section takes the readers through issues that the researchers should take note of when dealing with non-humans or animals.

The first step is for the researchers to write proposals justifying the need for them to do experiments with animals. The proposals should highlight the objectives of the study, risks involved and methods of handling such risks, methodologies involved and the research team members' profiles. The research proposal should be submitted to authorities responsible for allowing or denying permission. They should convince the authorities that the project (with animals) is beneficial to the communities and individuals involved. They should justify their experiment methods. They need to read literature related to their studies. Experimental methods that are not tried and tested should be avoided. The researchers should only rely on approved experiment methods.

Another key area is on the housing of animal subjects during the period of undertaking projects. Suitable housing for animal subjects should be provided and the researchers should work with animal psychologists throughout the project. The animal housing should be located away from noise locations. The human wildlife conflict should be avoided at all costs. Extra care is needed when researchers are dealing with endangered species. The researchers should study all the rules and regulations connected to endangered species.

The researchers should take good care of animals involved throughout the project. The acquired animals should be transported to the research site according to the rules. The research team should arrange everything from transportation to food. Animals should be given pain killers during the experiments. Exposing animals to pain is against the rules of research ethics. Researchers should work with animal experts in order to avoid violations of animal rights. These animal experts know all the strategies of handling animals and it should be a great mistake for researchers to work without animal specialists.

There are several ethics that are considered at the stage of research problem identification. Researchers should consider the discussion on intellectual property, how they deal with sensitive areas of research and children as well as the strategies of dealing with non-humans. They should work towards identifying risks and handling them in order to avoid harm. However, the ethics related to the selection of the research problems are not limited to the ones discussed above alone; researchers are requested to follow necessary ethics.

Ethics for Data Collection

There are ethics to consider for qualitative researchers before being involved in an intensive fieldwork. The qualitative researchers undertake their projects in locations called research sites. According to the dictates of social constructionist philosophy, research sites are considered to be holy places where data for research is gathered. Therefore, the researchers should respect research sites and avoid vandalizing these areas. They should refrain from compromising the status of the research sites and respect the locals found on such sites. Undermining the people living on research sites is not acceptable and researchers should consider the type of dressing to put on when undertaking fieldwork. They should not offend people living within the research site and abide with the local culture.

Permission to Undertake Fieldwork

The first step prior to the collection of data is for the researchers to request for permission to undertake their studies in the research sites from appropriate authorities and organisations involved with the study. The researchers are encouraged to avoid verbal agreements. It is always good to have written agreements. The permission may be denied or accepted and in the event of being denied, the researchers should refrain from the research sites. Permission is obtained at different levels i.e. from the authorities in charge of research sites and the individuals participating in the study (Creswell & Clarke, 2007).

After the permission has been granted, the researchers should move on to select gatekeepers. Gatekeepers are the people in charge of the research sites and they can help the researchers to familiarise with the research sites. They can also lead the researchers to the research subjects within the research sites (Guth, 2013). There is a lot for the researchers to learn from the gatekeepers and they should get advice about the research sites from them. There may be also need for the researchers to reciprocate by providing gatekeepers with some gifts but should not spoil them (Madondo, 2016b).

The gatekeepers will help the researchers to identify key informants. For a qualitative study to be successful, the researchers should get information from the right people who are called key informants in qualitative studies. The data is gathered from key informants and the research maybe invalid if the researchers fail to get information of research project from the right people. Therefore, the researchers should work closely with the gatekeepers in order to identify the right people who can provide reliable information about the project.

Informed Consent Signing

Informed consent is always sealed by the signatures of both the researchers and the research subjects. Informed consent may be verbal if the participants are not literate (Creswell, 2014). However, if the research subjects are able to read and write, the informed consent should be written. The informed consent form should cover the following; names of researchers, sponsors' details, information about the selection of research subjects, purpose, risks involved, strategies of dealing with risks, guarantee of confidentiality, assurance that research subjects are free to withdraw at any given time, contacts in case of issues to be clarified and a section for signatures. (Sarantakos, 2005)

An informed consent form is a legal document that can be produced in case of legal disagreements. The research subjects should be educated about the importance of such kinds of agreements if they are ignorant about their content. The research subject should be free to accept or deny participating in the study. Consent agreements are also needed when dealing with children and may be signed between the researchers and the people in charge of the children. Informed consent is a must if researchers are experimenting with humans. There will be no way the researchers can avoid them.

Photographing

In some cases, researchers may want to use cameras or phones to take photographs during the period of data collection. The researchers should get permission to take such photographs. If the permission to take photographs is denied, the researchers should look for other ways of getting data. The researchers should not take this for granted because it may cause a lot of inconveniences.

The same applies when dealing with children, researchers are not allowed to take photographs of children without consenting with the people in charge of them. Consent for photographing is needed when dealing with humans and non-humans and researchers should not take this for granted in order to avoid unnecessary confrontations.

Interviewing/Focus Group Discussion

The researchers should make sure that the research subjects within the research sites are not stressed during the interviews. The research subjects should feel at home and they should be motivated to continue with interviews. The interviews or the focus group discussions should be structured in such a way that the research subjects will be motivated. The researchers should avoid stress interviews.

The researchers should be also careful when recording the interview data. All the data recording equipment should be backed in order to avoid the loss of data. All in all, the researchers should work close to their gatekeepers in order to avoid inconveniences and probable risks should be eliminated before they cause harm to the participants.

Ethics Related to Data Analysis, Report Writing and Dissemination of Outputs

After the data has been collected, the researchers should embark on data analysis following the rules of qualitative paradigm. The researchers will code and categorize data for the purpose of writing the final reports. There are some ethics that should be observed and they are observed in order to protect the privacy of the research respondents. They should make sure that the following ethical rules highlighted below are observed.

Plagiarism

Plagiarism is defined as the copying and pasting of someone's ideas or work. Apart from copy and pasting, some cruel persons claiming to be researchers may ask other people to do projects for them. Plagiarism is not only evil but also an academic crime and disorder. There are some research companies established for the purpose of undertaking research projects on behalf of learners, such evil companies should be condemned by serious researchers.

Plagiarism challenges can be solved by understanding the popular American Psychological Association (APA) or any other styles of writing theses or projects' reports. There are rules that should be adhered to. Chapter 10 of this book contains detailed information on strategies of dealing with plagiarism. The college students should do their work on their own and should avoid engaging other persons or bogus research organisations. There are prominent people in society who were embarrassed when plagiarism was observed in their college theses/dissertations. Some of them were forced to resign from their national duties. This should be a wakeup call for college students and learners. Plagiarism is not different from snatching someone's wife and claim to the husband. It is not only an academic disorder/crime but also unethical and evil.

Confidentiality

The actual names of the research subjects should be concealed (Giordano, O'Reilly, Taylor & Dogra, 2007). The practise of exposing individual's names should be avoided at all costs. It is not only the names of individuals that should be concealed but also the names of organisations under the project. However, concealing of the names of the organisations should be done after a careful analysis of the sensitivity of the study. The research output should not be used to taint the images of the organisations under the projects.

There are several ways in which the names of individuals and organisations under the study may be concealed. The researchers may opt for the use of pseudo names or codes as well as numerals to represent real names of respondents who will have participated in a project. Concealing names is done in order to guarantee privacy and confidentiality of the research participants.

The research participants should be assured of confidentiality just before the process of gathering data. The research reports should not be used by security agents to punish respondents.

The Rule of Truths

The researchers should avoid the manipulation of data to suit their own interests. The data should speak and there should not be any form of massaging of information. The researchers should use genuine and acceptable methods of data analysis. The researchers should work as a team during the analysis and this is done in order to avoid bias. Triangulation of data may be done when validating the data.

Feedback to the Respondents

The researchers should report their findings from the data analysis to the research subjects before embarking on report writing. They have to go back to the research sites and allow the participants to receive feedback and approve it. In some cases, feedback given to the respondents may be used to validate the data. The researchers will also be able to make corrections during verification. The end of the verification process means the process of writing the report may begin.

Destruction of Research Data from the Sites

The researchers should keep information obtained from the research sites safely. The information should be protected with passwords which should be strictly guarded by the lead researchers. The information may be used anytime if there is need. However, there is controversy of the duration of keeping such data before destroying it. According to Sieber (1998), the information may be destroyed after the lapsing of five years or more. The author of this book recommends the information to be destroyed after one year.

The destruction of information depends on the type of the study. There are some studies that are done for the purpose of accomplishing undergraduate, master or PhD programs, the destruction of information gathered from research sites may be done soon after accomplishing their tasks or graduations. It is not healthy to keep information about the respondents for a long time. The longer the duration of storage the more the data is exposed to leakages, which is why it is advised to destroy the data soon after accomplishing the tasks.

Ethics for Report Writing & Dissemination of Research Outputs

Compiling research reports is sometimes difficult if the people compiling are not experts in the area of research writing. Extra care should be given on the linguistics styles and the jargon to be used. The researchers compiling should avoid language that provoke readers. Reports should just be written in simple and straight forward language. Careful use of language may neutralise emotions of the readers and the people directly affected by the studies. The use of radical and improper language may cause animosity and disturbances. The researchers should consider going for training before embarking on report writing. They may also consider writing as a team and try to send their reports to experts of research report writing for evaluation before publishing the results.

The researchers compiling reports should avoid gender biases and provoking religions and some sections of the society. The research reports should be free from any kinds of biases that my provoke those involved. Such kinds of biases may affect the credibility of the research outputs. There is therefore need for researchers to follow all ethics prescribed by qualitative research specialists.

Bracketing of researchers' feelings and emotions should be done during the report writing. The data from the fieldwork should speak and the emotions of the researchers should be supressed. It is unethical for researchers to supress the truths in their reports. There is need for verification of data and proofreading of the reports by fellow researchers and external experts. The research outputs should be accepted by the consumers of the research outputs. What is needed in report writing is objectivity and truth.

After the researchers are satisfied with the research reports, the owners or funders of the research studies may disseminate the information for public consumption. The researchers should allow the owners of the study to publish their results and they are not allowed to disseminate those results. The dissemination of results may be done via several ways. Some research outputs are disseminated via peer reviewed academic journals while other research outputs are disseminated for consumption via technology i.e. radios, television, newspapers and others. The dissemination of academic outputs is usually done through peer reviewed journals.

The research output is vital for development. The output is helpful to policy makers that is the reason researchers are encouraged to publish their work. They should publish their research outputs via proper channels in order to make sure that the information will reach its intended destinations. The researchers should make sure that principle research ethics are followed. Adherence to research ethics will boost the credibility of the study.

Critical Thinking Challenge

1. Suggest and justify ethical issues involved when dealing with ethnic communities?
2. Demonstrate your understanding of research ethics connected to non-humans?

Practical Experiences of Handling Ethics in Ethnographic Projects

This section contains information on how ethnographers are handling and applying ethics on the research sites. The article by Madondo (2016b) is published in this textbook with permission from the author and the publishing journal. The article highlights challenges that ethnographers are likely to face during the fieldwork and solutions are suggested. The study was undertaken with a sample of experienced university professors and PhD/master students who undertook ethnographic projects in Maasai and Hadzabe tribes of Northern Tanzania.

This empirical literature is very helpful to readers because it exposes the practical experiences of ethnographers when dealing with ethics before, during and after the field work. A phenomenological design was adopted to explore the lived experiences of researchers when undertaking their project in ethnic communities of Masaai and Hadzabe in Northern Tanzania. The qualitative project was under taken with 8 respondents. Four of them were university professors with significant experiences of undertaking qualitative projects within the research sites of Masaai and Hadzabe in Northern Tanzania. Three of them were university students who did their master theses/projects with Masaai and Hadzabe communities. The last responded was a PhD student who studied his thesis within these communities.

The project performed open, axial and selective coding outside the grounded theory. The readers will also be able to assess how it was done. Research ethics was found to be a core issue to consider when dealing with ethnic communities. The respondents share their practical experiences of handling and selecting gatekeepers and key informants. Selection of gatekeepers and key informants may be a challenge if not handled with care.

The researchers showed how they handled the reciprocity strategy during fieldwork. Some used to give cigarettes to the respondents and other incentives to ensure trust and respect. Giving incentives to the respondents is not a crime in qualitative studies. However, the incentives should not spoil the respondents. Spoiling respondents may be bad because it may lead to the provision of incorrect data. The readers should be able to experience how the ethnographers are handling ethical issues in their projects.

The readers should be also able to be psychologically touched by the empirical study. There are genuine cases of deaths reported. An undergraduate student from a local university missed death by a whisker when the community members tried to attack her. She was saved by the community elders and some members of a local political party. A clear message should be delivered to the readers that ethical issues should not be ignored in qualitative research projects.

All in all, the readers should be able to gain practical skills of handling ethics in ethnographic projects. They should also be able to be emotionally affected by the end of this chapter. The emotional experience will remind them to be careful when dealing with ethnic groups in future. I hope you will enjoy the reading.

Critical Thinking Challenge

By the end of the chapter, you should be able to answers the following questions;

1. You have been asked to gather data from an ethnic community X. State the ethical strategies that you would consider before undertaking the project?
2. How would you apply in your project the ethical strategies stated in question number 1?
3. After reading through the chapter, were you emotionally affected? In what ways
4. What have you learnt from the article?

Experiences in Conducting Ethnographic Studies in Marginalized Communities. A Case of Maasai and Hadzabe Tribes in Northern Tanzania

Silas Memory Madondo

Silas Memory Madondo is a mixed methods researcher based in East Africa silasethan@gmail.com

Introduction

A study was conducted in Maasai and Hadzabe communities in Northern Tanzania. The purpose of the study was to examine the experiences of selected eight ethnographers who carried out fieldwork in Maasai and Hadzabe communities. The research adopted a phenomenology design and structured interviews were used to collect data. 0pen, axial and selective coding were used to analyse the transcribed interview data (outside the grounded theory). It was observed that ethnographic studies which violate research ethics expose researchers and informants to risks. The risks ranged from negative attitudes to bloodshed. Ethnographers are encouraged to seriously consider ethical issues when planning for ethnographic fieldwork.

Conducting ethnographic studies may be dangerous if not handled with care. Sunday, October 2, 2016 was a dark day in the history of research. Three researchers of Selian Agricultural Research Institute were killed by Mvumi ward villagers in Dodoma, Tanzania. Teddy Lumanga, Faraji Marufu and their driver Nicas Magazine lost their lives. The villagers were not informed in advance about the presence of the researchers in their backyard (Ubwani 2016). The researchers failed to address the rules of research ethics properly. Researchers should be informed about the dangers associated with conducting ethnographic studies in marginalized communities in order to minimize and avoid risks. Figure 1 below shows the scene of the accident.

Overview of Hadzabe and Maasai Tribes

Maasai and Hadzabe (Hadza) communities are located in the northern parts of Tanzania. Maasai and Hadza tribes are the only tribes in East Africa that are still relying on traditional way of life. According to the advice from Feinstein (2008), Maasai people live a semi-nomadic lifestyle. They are pastoralists with large heads of cattle and they migrate their animals through

Ngorongoro crater in search of pastures. Maasai community is a patriarchal society practicing polygamy and women are always submissive to their men. They are not involved in decision making on key societal issues. The Maasai community is not seriously participating in politics and economy of Tanzania mainland.

Figure 1. Presents the remains of the torched vehicle of the deceased researchers
Source: The citizen newspaper (Tanzania) Sunday October 2016

Figure 2. Presents a photograph of Maasai warrior
Source: Paul Caridad, Sept 2014

Maasai and Hadza communities differ on livestock keeping. Hadzabe people do not keep livestock, according to the National Geographic publication of November 2009, Hadza people are hunters and gatherers. They live in the forests with animals though there are some current changes caused by globalization. Hadza people live around Lake Eyasi in the central rift valley. The total population of Hadza people is estimated at 1000.

Figure 3. Presents a photograph taken from Hadzabe
Source: National Geographic, November 2009

Purpose of the Study

The purpose of the study was to examine the experiences of ethnographic researchers in marginalized areas (Maasai and Hadzabe in Northern parts of Tanzania). The research answered the following questions,

- What are the key issues to consider when conducting ethnographic studies in Maasai and Hadza communities?
- What are the challenges facing ethnographic researchers when conducting studies in Maasai and Hadza communities?

What is Ethnography Research?

Researchers proposed several definitions for ethnography design. According to Angrosino (2007) as cited by Sangasubana (2011), ethnographic designs study the patterns of human lived experiences. Creswell and Clarke (2007), defined ethnography as research projects which describe and interpret culture/social groups or a system. Ethnographic studies examine the cultural characteristics of a certain group of people (Madondo, 2016). Guth (2013) defined ethnography as '....unobtrusive research through observation and limited interaction' with the informants. Limited interaction with informants helps to reduce bias in data collection.

Ethnographic research involves extensive ground work called fieldwork. It involves the study of people's languages, behaviours and artefacts. Ethnographers should ask for, what people do? (behaviour), what they say (language) and what they make and use? (artefacts) (Creswell & Clarke, 2007). Ethnography design may be used to study the marginalized groups (Wolcott 1999 as cited by Sangasubana, 2015). Maasai and Hadza communities are classified under marginalized communities because of their slow speed of adapting to modernization.

Origin of Ethnography Research

This section is based on advice from Ellen (1984). The year 1840 was important in the history of ethnographic studies; anthropology was introduced as a branch. The author went on say that an American named Henry Schoolcraft was the first person to introduce ethnographic information and the Bureau of Ethnology was introduced in 1879 to study Indians. Participant observation is an important method for ethnographic research. Participant observation is said to have been coined by Malinowski in 1914. He did a prominent fieldwork among Trobriand Islanders of New Guinea.

Core Issues to Consider when Conducting Ethnographic Studies

Handling an ethnographic study is complicated if a researcher is not a participant observer, (Ellen, 1984). Madondo (2016) highlighted that participant observation is central to ethnographic study. However, participant observers may find it difficult to control their preconceived ideas about the communities that they are familiar with, this may be the source of bias to the study. Ethnographic studies can be conducted by non-participant observers as well but they (non-participant observers) should be exposed to the culture, politics, language and the way of life of the communities under the study. Ethnographic researchers should understand the key issues to consider when conducting studies with communities because there are several risks involved.

Ethnographic studies are more concerned about ethics and ethnographers should get clearance from the responsible authorities before conducting their studies. Pontzer et al (2012) did a study in Hadza community, their aim was to make a comparison of energy consumption and obesity between Hadza and the western lifestyle. They got approval from Washington University Institutional Review Board, Tanzania Institute for Medical Research and Commission for Science and Technology of Tanzania.

Ethics in ethnographic studies do not end with permission, consent agreement should be signed by participants. In areas where literacy levels are too low, verbal agreement may be acceptable to employ. Pontzer et al (2012)'s study in Hadza community was based on verbal consent because of low literacy rates. It is therefore advisable for ethnographic researchers to include a statement of ethics in their reports.

Knowledge of the selection of gatekeepers is another key area to consider. Sangasubana (2011) believes that the selection of gatekeepers should be done just before the collection of data. Gatekeepers according to Creswell and Clarke (2007) are people who help the researcher to get access to the research site. In most cases, the gatekeepers are community leaders and their duty is to facilitate peaceful entry of ethnographers to the research sites. The researchers should look for right the people who can link them to the gatekeepers.

Gatekeepers may also link the researchers to the key informants. Creswell and Clarke (2007) defined key informants as people who provide useful information to the study. Guth (2013) argued that ethnographers should start with a game plan. They should know the key informants and the kinds of data to be collected. Data collected from key informants is always valid and the aim of ethnographic studies is to get valid data which is reliable.

Sangasubana (2011) talked about validity and reliability of data as important components of ethnographic studies. He defined reliability as the internal and external consistence of the research findings. The researcher does not agree with Sangasubana (2011)'s view of validity and reliability of a qualitative study. Validity and reliability are terms used in a quantitative study and in a qualitative study (like ethnography) researchers talk of trustworthiness of the research data. According to Shenton (2004) trustworthiness is the credibility, transferability, dependability and conformability of the research findings. However, trustworthiness is the equivalence of reliability and validity in qualitative studies.

There are several ways of improving the trustworthiness of a qualitative study. According to Shenton (2004), randomization can be used during the process of selecting key informants in a study and this will help to reduce bias. Qualitative research findings according to Shenton (2004) can be generalized (transferability) to specific similar contexts but not to larger population. The researcher should provide information about the delimitations of the study, the nature of the respondents and study in order for other readers to generalize the research findings to other contexts. Triangulation of methods, designs and approaches can be used to boost the credibility of the research study as well.

Madondo (2016) talked about the reciprocity in ethnographic studies. Reciprocity is a symbiotic relationship between the researcher and the researched in a study. The study should benefit both the researcher and the researched. Creswell and Clarke (2007) support the idea of reciprocity. The researcher should provide the informants with some incentives i.e. lunch/ drinks, cash for transport and cigarettes depending with the nature of the study and informants. However, the researcher should not spoil the informants because they may be induced to provide misleading information.

Guth (2013) emphasized that ethnographers should approach the study with an open-mind and fresh eye. The researchers are encouraged to start by 'bracketing' their preconceived ideas about the problem under the study (Creswell & Clarke, 2007). Bracketing is not an easy task and it is psychological. Ethnographers are encouraged to observe the ethics of conflict of interest depending with circumstances.

Guth (2013) argued that ethnographic studies involve complex data collection and coding. In most case, ethnographic studies just like grounded theory need researchers to have multiple visits to the research site. Ethnographers should therefore plan carefully for data collection. Note pads, cameras and recorders are needed during data collection.

Challenges Associated with Ethnographic Studies

Dealing with people's traditions and culture is not an easy task and may expose both the researcher and informants to risks if not handled properly. Madondo (2016) supported the fact that the duty of the researcher is to make risk assessment before the commencement of the study. Failure to make a risk assessment may expose both the researcher and the researched to threats. There are several risks involved in conducting an ethnographic study.

The researchers and key informants may be under the threat of reactivity. Sangasubana (2011) defined reactivity as an effect of the presence of the researchers to the subjects on the research sites. The members of the community may be annoyed by the presence of the researchers and become violent. Ethnographers are therefore encouraged to handle ethics properly and look for advice from the gatekeepers. The key informants may be also under the threat from the researchers. Some researchers may feel negative towards key informants (Lopez-Dicastillo & Belintxon, 2014).

Bracketing of the researchers's preconceived ideas about the study is a problem especially to participant observers (Lopez-Dicastillo & Belintxon, 2014). It is always difficult to control psychological phenomena like bracketing but ethnographers are encouraged to be fair because biasness is a threat to the credibility and transferability of the study findings.

The other challenge of ethnographic study is on reliability and validity of data (Sangasubana 2011). It is always difficult for the researcher to come up with credible and dependable findings if participants are not key informants. Some key informants are not free to disclose their traditional and cultural values that are considered sacred by their societies. Under such circumstances, the participant observer may help to screen and validate data from informants.

Lopez-Dicastillo and Belintxon (2014) argued that ethnographers need to be natives. What does this mean? Are non-participant observers not eligible to conduct ethnographic studies? Certainly no, anyone can conduct ethnographic studies but should be exposed to the rules and guidelines of conducting them. Participant observers just like non-participant observers have their own weaknesses and ethnographic studies can be done by anyone.

Method

The study adopted a qualitative approach and the research design used was phenomenology. Phenomenology is used to study the livid experiences of several individuals (Creswell & Clarke 2007). The study collected information from 8 ethnographers that are experienced in conducting studies in Maasai and Hadza communities. The study managed to examine their lives experiences of conducting researches in Maasai and Hadza communities. Each and every participant was allowed a 20 minutes interview. The researcher exercised extreme caution by minimizing interference during the interviews.

The first step in data analysis was the manual intelligent verbatim transcription of the data from audio recordings. The researcher analysed the data through performing open, axial and selective coding outside the grounded theory. The process of coding was done manually as well, and the coding was done basing on cases and nodes.

3 FINDINGS AND DISCUSSION

Demographic Characteristics of Respondents

The study was based on the information provided by 8 ethnographers who has experiences of conducting studies in Maasai and Hadza communities. Out of the 8 respondents, 2 were anthropologists teaching in public universities in Tanzania, the other 2 were historians, one was a lecturer of rural and urban geography at university level and another respondent was a university lecturer of climate change. The study also took on board one political scientist and a community developer. All these respondents had experience of conducting studies in Maasai and Hadza communities. Figure 4 below presents the demographic characteristics of respondents.

The analysis of data was done basing on open, axial and selective coding outside the grounded theory and this section presents the findings and discussion of the study in accordance with the research questions.

Core Issues to Consider when Conducting an Ethnographic Study

The disintegration and reassembling of the transcribed data managed to identify 5 major categories from the data. The categories are research ethics (RE-Q1), gatekeepers and key informants (GKI-Q1), validity and reliability (VR-Q1), reciprocity (RP-Q1) and data collection (DT-Q1) related category. The categories represent key issues to consider when conducting ethnographic studies according to the selected 8 researchers with various experiences of conducting ethnographic studies in Hadza and Maasai communities.

Figure 4. Presents the demographic characteristics of the respondents
Source: Researcher 2016

Participant	Area of Study	Period	Profession
Researcher 1	History	2015	Master Student for History (Hadza area)
Researcher 2	Anthropology	1995-2000, 2005 & 2014	University Professor (Hadza & Maasai)
Researcher 3	Climate Change	2005 & 2016	PhD Student (Maasai)
Researcher 4	Political Studies	2014	University lecturer (Maasai)
Researcher 5	History	2016	Master Student (Hadza)
Researcher 6	Rural &Urban Geography	2009, 2015	University Lecturer (Maasai)
Researcher 7	Anthropology	2010 to present	University Senior Lecturer (Hadza & Maasai)
Researcher 8	Community Development	2016	Master Student (Maasai)

Validity and Reliability/Trustworthiness (VR-Q1)

This category was found to be the core category basing on selective coding. It was observed that the researchers validated and ensured the credibility of their data by triangulating instruments of data collection. The five researchers used several methods of data collection in order to validate the data. Researcher number 8 claimed that he visited the research site 3 times in order to validate the data this is what he had to say,

'I was forced to visit the research site 3 times because the data that I collected had some contradictions ....'

All researchers claimed that they validated the data by making comparisons with other related studies done before and that was done through literature search. Sangasubana (2011) agreed that reliability and validity are core elements on an ethnographic study. Shenton (2004) suggested triangulation of instrument as a key method of validating research data.

Research Ethics (RE-Q1)

This was another key category identified and it is connected to the core category (reliability and validity) in the sense that the aim of research ethics is to improve the credibility and dependability of the research findings. Out of 8 respondents, 6 got permission from Commission for Science and Technology (COSTECH) in Tanzania and the other 2 respondents got permission from Pale Anthropology Field School at Olduvai Gorge. Consent agreement is central to ethnographic studies and all the 8 respondents had verbal consent with informants because of literacy problems. Researchers number 2 & 6 claimed that they had both written and verbal consent. Pontzer et al (2012) only used verbal consent when they conducted their study in Hadza area.

Gatekeepers and Key Informants (GKI-Q1)

Ethnographers should consider gatekeepers and key informants when conducting their studies. Researcher number 2 & 6 were introduced to the gatekeepers and key informants by local NGOs working in the area. Researcher number one was introduced to gatekeepers and informants by the local leadership of the ruling party Chama Cha Mapinduzi (CCM) during the 2015 elections. This is what he had to say,

' ......and I took advantage of the campaigning, CCM campaigning team in the Hadzabe area introduced me to the participants.....'

Researcher number 8 who was a participant observer used snowball sampling to locate key informants with the help of the gatekeepers. Guth (2013) stated that ethnographers should start by identifying key informants of the study.

Reciprocity (RP-Q1)

The key function of reciprocity is to induce the informants to provide credible and dependable information. The main aim of reciprocity is to ensure that the study is valid and reliable and this is how reciprocity is linked to the core category (validity and reliability). Researcher number 1 used to give cigarettes to informants and he said that majority of informants in Hadza were smokers. He also said that tourists are loved by Hadza people because they bring them second hand clothes. Two researchers said they used to pay gatekeepers for their service, but 5 of them claimed that they never reciprocated. Creswell and Clarke (2007) supported the idea of reciprocity as an effective tool that may be used to induce informants to provide data.

Data collection procedures (DC-Q1)

All researchers considered carefully the processes of data collection. The researchers used note pads to write key information from the field. Cameras and mobile phone were used for photographing. Researchers number 1,3 & 6 used iPhone to record important information and the data was backed by iCloud. The process of data collection was designed to support the core category (validity and reliability). Perfect data collection aims at improving the trustworthiness of the study.

Challenges Associated with Ethnographic Studies

The disintegration and reassembling of the transcribed data on challenges associated with ethnographic studies in Maasai and Hadza communities produced 5 categories. The categories of challenges associated with ethnographic study are, reactivity (RC-O2), bracketing and bias (BB-Q2), validity and reliability of data (VR-Q2) or trustworthiness, participant and non-participant challenges (PP-Q2) and data collection (DC-Q2) related challenges.

Trustworthiness of the Research Findings/ Validity and Reliability (VR-Q2)

This was the key/core category of the challenges affecting ethnographers when conducting studies in Maasai and Hadza communities. All the challenges affecting the ethnographic studies have impact on the credibility and dependability of the research findings. It was strange to observe that no single researcher managed to send back the findings of the study to the key informants for verification and validation. Researchers are always advised and encouraged to give feedback to the informants for the purpose of verification. Two of the researchers complained that a significant number of informants were supplying irrelevant information and that resulted in vigorous data cleaning before data analysis. Madondo (2016) stated that validation of data by giving feedback to informants is important.

Figure 5. Presents a summary of open, axial & selective coding for issues to consider when undertaking ethnographic studies in Maasai & Hadzabe communities
Source: Researcher 2016

Open Coding	Axial Coding	Selective Coding
Research Ethics (RE-Q1) • Got written permission from COSTECH and Pale anthropological Field School at Olduvai Gorge • Consent was based on Verbal agreement	Research approval	Validity and reliability of research data
Gatekeeper and Key Informants (GKI-Q1) • Introduced by NGOs working in Hadza • Connected to gatekeepers by ruling party CCM during campaigning • Gatekeepers helped to link informants • Introduction was easy because the researcher was from the same tribe	Avenues for linking gatekeepers & informants	
Validity and Reliability (VR-Q1) • Triangulation of instruments • Visiting research site more than once • Validation of the report findings by participants • Validating by other related studies	Credibility and conformability of the data	
Reciprocity (RP-Q1) • Gave cigarettes • Paying gatekeepers for the service • Tourists gave second hand clothes • Some never reciprocated	Motivation and research tips	
Data collection (DC-Q1) • Note pads used • Cameras used • Tape recorders used	Instruments of research	

Reactivity (RC-Q2)

Conducting research about culture and tribes may be sometimes dangerous if not properly handled. Researcher number 1 claimed that a student from a local university was attacked when conducting her study in Hadza area. She escaped following the intervention of a local district leader. Researcher number 6 said that the environment was strange during the beginning of the study but his attitude improved with time. Researcher Number 8 admitted that he was sometimes having negative feelings towards some of the informants. Lopez-Dicastillo and Belintxon (2014) acknowledged that researchers may have a challenge of feeling negative towards some informants.

Bracketing and Bias (BB-Q2)

This was another category of challenges observed. Researcher number 8 who was a participant observer admitted that his knowledge of Maasai culture had a little impact on some of his decisions and conclusions during the research process. Bracketing of researcher's interests is sometimes difficult for participant observers. Out of all the 8 researchers, none used randomization to pick informants, Shenton (2004) argued that randomization can be used to reduce bias in research.

Data Collection Challenges (DC-Q2)

There are several data collection challenges associated with ethnographic studies. The researchers complained of huge data which was demanding during the process of transcription. Researcher number 6 admitted that he took 2 months to transcribe the data and the data was huge. All the 8 researchers analysed the data manually and there was none who had the knowledge of qualitative data analysis software. Two of the 8 researchers complained of the missing data.

Participant and non-Participant Challenges (PP-Q2)

Lopez-Dicastillo and Belintxon (2014) argued that ethnographers need to be natives. However, both participants and non-participants have their challenges. Participant observers had problems related to bracketing and admitted that their knowledge of Maasai had a little contribution to their judgments and decisions during the research process. The non-participant observers faced problems of communication language and some complained that the environment was strange in the beginning of their studies.

Figure 6. Presents a summary of open, axial and selective coding for challenges facing ethnographers undertaking studies in Maasai & Hadzabe communities
Source: Researcher 2016

Open Coding	Axial Coding	Selective Coding
Reactivity (RC-Q2) • Fear of the strange environment • University student survived attach from Hadza people • Poor cooperation during interviews • Negative attitude towards some informants	Non responsive & hostility	Trustworthiness of data challenges
Bracketing and Biasness (BB-Q2) • The situation did not allow randomization • Heavy interference of researchers during interviews • Knowledge of the tribes had a little influence to the study	Preemptive information and bias	
Validity and Reliability (VR-Q2) • No validation of research findings in some cases • Some information provided was irrelevant	Trustworthines s credentials	
Participant and non-Participant Challenges (PP-Q2) • Language problems for non-participant observers • Strange environment for non-participant observers **Data Collection (DC-Q2)** • The data was huge • No knowledge of software for qualitative data analysis s • Difficulty in analyzing data manually • Transcription of huge data done manually • Problem of missing data	Communicatio n & environmental barriers Bulky data software Incompetence	

4 CONCLUSION

According to the outcome of open, axial, and selecting coding, the theme of validity and reliability was found to be the core category of the challenges affecting ethnographers in Maasai and Hadza communities. Validity and reliability category was also found to be the core category of the factors to consider when conducting ethnographic studies in Maasai and Hadza communities. The study also observed that researchers had little information on qualitative data analysis techniques and computer software for qualitative data analysis.

ACKNOWLEDGMENT

I want to express my gratitude to Mount Meru University for sponsoring this project.

REFERENCES

Andre, C., & Velasquez, M. (2015). *Of Cures and Creatures Great and Small.* Academic Press.

Bandura, A. (1979). Principles of Behaviour Modification. New York: Holt, Renehart and Winston.

Creswell, J. W. (2013). *Qualitative Inquiry and Research Design: Choosing among Five Approaches* (3rd ed.). Sage.

Creswell, J. W. (2014). *Research Designs: Qualitative, Quantitative & Mixed Methods Approaches*. SAGE Publications, Inc.

Creswell, J. W., & Clark, P. (2007). *Planning and Conducting Mixed Methods Research* (2nd ed.). Sage.

Creswell, J. W., & Plano Clark, V. L. (2011). Designing and conducting mixed methods Research (5th ed.). Thousand Oaks, CA: Sage.

Ellen, R. F. (Ed.). (1984). *Ethnographic Research: A Guide to General Conduct*. Academic Press London.

Feinstein, S. A. (2008). *Research Study on Individuals with Disabilities in the Maasai Tribe in Tanzania* (Published Thesis). Augustana College.

Giordano, J., O'Reilly, M., Taylor, H., & Dogra, N. (2007). Confidentiality and Autonomy: The Challenge(s) of Offering Research Participants a Choice of Disclosing their Identity. *Qualitative Health Research*, *17*(2), 264–275. doi:10.1177/1049732306297884 PMID:17220396

Guth, D. W. (2013). *A Simple Guide to Ethnography*. http://www.dguth.journalism.ku.edu/Ethnography.html

Lincoln, Y. S., & Guba, E. G. (1985). *Naturalistic inquiry*. Sage. doi:10.1016/0147-1767(85)90062-8

Lopez-Dicastillo, O., & Belintxon, M. (2014). The Challenges of Participant Observer of Cultural encounters with an Ethnographic Study. *Procedia: Social and Behavioral Sciences*, *132*, 522–526. doi:10.1016/j.sbspro.2014.04.347

Madondo, S. M. (2016a). Essentials of Social Science Research. Mount Meru University Publications Unit.

Madondo, S.M. (2016b). Experiences in Conducting Ethnographic Studies in Marginalized Communities. A Case of Maasai and Hadzabe Tribes in Northern Tanzania. *Imperial Journal of Interdisciplinary Research, 2*(12).

Milgram, S. (1974). Behavioral Study of Obedience. *Journal of Abnormal and Social Psychology*, 1963.

Payne, G., & Payne, J. (2004). *Key Concepts in Social Research*. Sage Publications. doi:10.4135/9781849209397

Pontzer, H., Raichlen, D. A., Wood, B. M., Mabulla, A. Z. P., Racette, S. B., & Marlowe, F. W. (2012). Hunter-Gatherer Energetics and Human Obesity. *PLoS One*, *7*(7), e40503. doi:10.1371/journal.pone.0040503 PMID:22848382

Sangasubana, N. (2011). How to Conduct Ethnography Research. *Qualitative Report*, *16*(2), 567–573.

Sarantakos, S. (2005). *Social Research* (3rd ed.). Palgrave Macmillan. doi:10.1007/978-0-230-20901-5

Shenton, A. K. (2004). Strategies for Ensuring Trustworthiness in Qualitative Research Projects. *Education for Information*, *22*(2), 63–75. doi:10.3233/EFI-2004-22201

Sieber, J. E. (1998). Planning ethically responsible research. In L. Bickman & D. J. Rog (Eds.), *Handbook of applied social research methods* (pp. 127–156). Sage.

Smith, D. (2009). Five Principles for Research Ethics. *Monitor Staff, 34*(1).

Ubwani, Z. (2016, Oct. 2). Shock over Slaying of Arusha Researchers. *The Citizen Newspaper*.

Chapter 2
Qualitative Research Philosophy

ABSTRACT

The chapter is divided into six sections. The first and second sections expose readers to the origin of social constructionist philosophy and the reasons behind its emergence respectively. The third section discusses the key tenets of the social constructionist philosophy based on ontology, epistemology, axiology, methodology, and rhetoric. A comparison of social constructionist and other methods is done in Section 4, and Section 5 addresses the unique characteristics of ideal qualitative projects. The chapter is concluded by a question and answer section where important questions about the research philosophy are suggested and answered.

1 KEY CONCEPTS

- **Research Philosophy** – positions taken in research that shape the way researches are undertaken. Worldview/paradigm/approach may be used in place of philosophy.
- **Qualitative Philosophy** – the approach/worldview/paradigm/philosophy for qualitative studies is called social constructionist.
- **Epistemology** – the acquisition of knowledge in research.
- **Ontology** – the nature of reality in research.
- **Axiology** – values in research.
- **Methodology** – the use of deductive and inductive methods in research.

DOI: 10.4018/978-1-7998-8549-8.ch002

- **Rhetoric** – the nature of language used in research reports.
- **Tenets** – core principles of a phenomenon.

2 LEARNING OUTCOMES

By the end of this chapter, readers should be able to;

- Trace the origin of qualitative research philosophy.
- Narrate the rationale behind the emergence of social constructionist.
- Describe the tenets of social constructionist philosophy.
- Make a comparison of social constructionist with other philosophies.
- Grasp key areas to address in ideal qualitative projects.

Case Study

David, a fourth year student at University of Zimbabwe planned for a qualitative project with Rozvi people staying in the Southern parts of Zimbabwe. In his project proposal, David claimed that his study would adopt a social constructionist approach. The study proposed to use three major methods of data collection; structured interview, document analysis and a likert scale. He also proposed to adopt random and convenience sampling to pick samples from the study population. David's proposal could not be approved by his supervisors, they were convinced that David proposed some methods for positivists in his proposal. By the end of reading this chapter you should be able to identify methods for positivists that were proposed by David.

3 INTRODUCTION

David had little knowledge of research philosophies that is why he proposed wrong methods for a social constructionist project. Research ideologies vary according to the nature of the study. The qualitative researchers subscribe to a paradigm different from quantitative specialists. There are several philosophies in research but there are three popular ones. Social constructionist, positivist and pragmatist are popular philosophies in research. Qualitative, quantitative and mixed methods studies are linked to social constructionist, positivist and pragmatist philosophies respectively. However, the key focus of this textbook is on discussing qualitative methods and data processing. This chapter is divided into the following sections;

- Genesis of Qualitative Research Philosophy
- Reasons for the Emergence of Social Constructionist Philosophy
- Traits of Social Constructionist Philosophy
- Tenets of Qualitative Research
- Key Areas to Address in Ideal Qualitative Projects
- Question and Answer Session

Genesis of Qualitative Research Philosophy

The term worldview can be used in place of research philosophy. The term was coined by Guba (1990). Another prominent term that can be used interchangeably with research philosophy is paradigm. Research paradigm can be traced to Thomas Kuhn who survived from 1922-96. The term 'research approach' can also be used in place of research philosophy. The way researchers undertake their studies is guided by research philosophy/paradigm. These philosophies vary in terms of their interpretations of epistemology, ontology, axiology, methodology and rhetoric.

Social constructionist philosophy is claimed to have originated in 1960s. However, positivist worldview is said to be the oldest philosophy which preceded the social constructionist philosophy (Borg & Gall, 1989). According Smith (1983), positivist philosophy originated during the 19th Century. Comte, Mill, Durkheim, Newton and Locke are founders of the positivist philosophy. It is therefore interesting to note that the social constructionist philosophy's origin was provoked by the weaknesses of the positivist paradigm. The social constructionists could not agree with the tenets of the positivists. The discord in worldview between the protagonists was labelled paradigm war and the dispute resulted in the emergence of social constructionist paradigm.

Research paradigm in projects is important because it determines the selection of research questions, nature of research instruments, research designs, data collection methods, sampling procedures and methods of data analysis. The aim of this textbook is to enable researchers to undertake purely (ideal) qualitative research projects that are devoid of quantitative traits. The readers should be able to grasp the qualitative methods as well as the processes of qualitative data analysis. The differences between qualitative and quantitative methods are based on paradigms that they subscribe to.

Reasons for the Emergence of Social Constructionist Philosophy

The disagreement between qualitative and quantitative protagonists resulted in a war of words. The disagreement became popularly known as the paradigm war. The two groups could not agree on epistemological, ontological, axiological, methodological and rhetorical worldviews. The paradigm war resulted in the constructionist paradigm in the 1960s. Some may be forced to subscribe to the perspective that the emergence of social constructionist paradigm was the rejection of the positivist philosophy. The adoption of pragmatism in mixed methods research has been trying to bring harmony between qualitative and quantitative research specialists.

An example below is a summary of an articles by Johnson and Onwuegbuzie (2004) 'Mixed Methods Research: A Paradigm whose time has Come'. The summary highlights the paradigm war between the quantitative and qualitative protagonists that led to the development of mixed methods research. The readers should acknowledge that it was the paradigm war that led to the development of social constructionist philosophy as well as the emergence of the third force now popularly known as mixed methods research.

The article exposed that the debate or war between quantitative and qualitative followers has been going on for some more than 100 years now. Quantitative researchers argued that social observations in research should be done the same way physical scientists are undertaking them and there should be a separation between the researcher and the researched. Moreover, the social science enquiry should be objective rather than being subjective.

Qualitative researchers hit back, with Guba (1990) suggesting that, mixing quantitative and qualitative approaches is impossible. Qualitative researchers rejected positivism philosophy and hold their position of the existence of multiple realities in research, and argued that, it is very difficult to separate the researcher from the researched. They further support the use of inductive approaches in research and rejected the passive way of writing in favour of the detailed one.

The article gave a detailed analysis of the similarities between qualitative and quantitative. It is because of these commonalities between the two paradigms that Johnson and Onwuegbuzie (2004) are advocating for the adoption of mixed method research (MMR). They argued that MMR will drive research to become inter disciplinary, dynamic and that MMR offers the best and comprehensive chance of answering research questions.

Both (quantitative and qualitative) paradigms seek to explain and describe data as well as making use of empirical observations to address research questions. Furthermore, the aim of both paradigms is to minimize confirmation bias that may affect the study and that both paradigms seek to research on human beings and environment that they live and evolve. It is because of these similarities that the article is advocating for the adoption of MMR as the third wave.

The paper highlighted the strengths and shortfalls of qualitative paradigms; qualitative paradigm studies smaller samples in detail and can be used to study in detail individual cases. Moreover, qualitative paradigm describes and explain complex phenomenon and is also good for cross-case comparisons. Qualitative research is responsive to the changes that occur during the study.

On the other hand, qualitative research has some shortfalls; its findings cannot be generalized to a larger population and the qualitative research findings are not popular with policy makers. The process of research in qualitative is tedious and researcher can easily influence the results.

Quantitative just like qualitative has its own strengths and weaknesses; it can be used for projects with larger samples and can be used to validate and test theories. Moreover the results of the quantitative paradigm can be replicated to different contexts and the process of collecting data is less demanding compared to qualitative.

On the negative side, the sample of a quantitative research may not be a reflection of what is on the ground if sampling procedures are not used wisely and the outcome of a quantitative study may not apply to unique contexts and individuals. Majority of quantitative theories may be irrelevant to contexts and quantitative research can be affected by confirmation bias.

The summary above highlights the discord between quantitative and qualitative researchers. The dispute known as paradigm war helps the readers to understand the reasons that led to the emergence of social constructionist paradigm. The social constructionists could not agree with the positivists' worldview on objective reality, generalizability of research findings to other contexts, the nature of instruments and methodology. That was the context in which social constructionist philosophy originated.

Critical Thinking Challenge

1. State reasons for the emergence of social constructionist approach?

Traits of Social Constructionist Philosophy

The social constructionist philosophy guides the way qualitative researchers undertake their studies. This section is a discussion of the key tenets of social constructionist philosophy based on epistemology, ontology, axiology, methodology and rhetoric. The section is based on the condensed views of Cherryholmes (1992), Murphy (1990), Patton (1990), Newman & Benz (1998), Crotty (1998), Phillips & Burbules (2000). Epistemology in the context of this textbook means knowledge generation and gathering in research. The researchers should know how knowledge is acquired and its meaning in research. The positivists and social constructionists could not agree on the worldview of knowledge in research.

Ontology means the nature of reality in research. Is the reality objective or subjective? Is there a reality in research or it is a work in progress? There positivists and social constructionist couldn't agree on ontology though there is a sort of an agreement among the social constructionists, post positivists and pragmatists. The positivists have a simplistic view of ontology.

Axiology means values in research. What values should be considered in research, individual/majority or majority values? The axiology also addresses issues related to research ethics. Adherence to research ethics has been a challenge to the positivists, which resulted in several violations of freedoms and rights of research subjects caused by the positivists. The positivists and social constructionists subscribe to different views on axiology.

Methodology means the methods of undertaking research studies. There has been a tense debate on the proper methods of undertaking research studies. The way quantitative and qualitative researchers undertake their research projects varies. The variation is because of their different worldview. Should researchers use inductive, deductive or both approaches? Which approach is the best? This should be addressed in this chapter.

Rhetoric means the use of language in research. The language used by the qualitative researchers differs from the language used by quantitative researchers. Should the researchers use formal or informal language in their research reports? The pragmatists believe in anything that works. There is no problem in using both informal and formal language provided it works.

The Epistemological Worldview of Social Constructionists

The social constructionists hold the belief that knowledge is difficult to get because it is multiple and fragile. Knowledge is constructed with people as they engage with the world and the knowledge is generated inductively. The positivists are holding a different view. They believe that knowledge is generated deductively and is easy to get. The view of the positivists was also challenged by the post positivists, they assume that there is no ultimate knowledge because knowledge is ever changing. The same position has been adopted by the pragmatists.

The social constructionists also accept that knowledge is subjective and created between the researcher and the researched. The implication is that the researcher and the researched cannot be separated. The researcher should interact with the respondents during the data gathering. It is very difficult for the researchers to administer interviews in absentia. On the other hand, the positivists believe in ultimate knowledge and that the researchers should not interact with the researched (respondents) during the gathering of data. They classify the researchers' interactions/interferences with the researched as bias. The implication is that researchers should not interact with the respondents so that biases can be reduced. The pragmatists support anything that works, the researcher may not interact (or may interact) with the researched provided it works (Farjoun, Ansell & Boin, 2015).

Ontological Worldview of Social Constructionists

Researchers differ on the way they view the nature of reality. The social constructionists challenged the positivists' position of the existence of a single reality that is in the universe waiting to be observed. They believe that there are multiple realities that are difficult to observe. Therefore, a qualitative project/study involves a vigorous fieldwork aiming at exposing complex realities. The post positivists are of the view that reality keeps on changing and it is therefore difficult for the researchers to get it. The pragmatists agrees with the view of social constructionists on the nature of reality. They believe that reality is ever-changing and sophisticated methods should be employed to explore realities.

According to the social constructionists, the nature of reality is subjective and generalizations of research findings to other contexts should be done with extra care. The research findings should only be generalized to other similar contexts and not to different contexts. On the other hand, the positivists have no problems with the generalizations of the research findings because they believe in the existence of ultimate reality. The post positivists believe that there is no need for researchers to generalise research findings because contexts are totally different.

Axiological Worldview of Social Constructionists

The social constructionists respect individual and minority values. The individual values/ideas are considered in study's findings and conclusions. The positivists have been challenged for ignoring the individual values. Their findings are only biased towards the majority views. The conclusions are valid if they are supported by the majority and the individual or minority views. The individual and minority values are ignored. According to Hancock (2002), qualitative research is concerned with feelings, experiences and opinions of individuals. In contrast, quantitative studies are concerned with what the majority say. What the majority supports is always right and the conclusions are based on the majority rule.

The implication to practice is that social constructionists' projects/studies work with smaller sample sizes. They are not concerned with larger sample sizes but the views, opinions and experiences of the respondents. For example a biography design of a qualitative project involves a study of a single person because the social constructionists are concerned with the individual values and opinions rather than larger sample sizes. The qualitative studies involve smaller sample sizes because they involve an intensive fieldwork that is time consuming (Hancock, 2002). Positivists (quantitative studies) work with larger sample sizes. A quantitative study is only reliable and credible if it is backed by a larger sample size. Bluman (2007) defined a lager sample as a sample with not less than 30 respondents. However, a definition of a larger sample size varies with contexts because some studies are undertaken with sample sizes of above 100 respondents.

Another aspect of axiology is on research ethics. The social constructionists respect ethics when undertaking qualitative projects/studies. They subscribe to the view that ethics in research should be adhered to by the researchers. The researchers should observe ethics prior and during the period of undertaking their studies as well as when they are publishing their outputs. They adhere to research ethics because they value individuals and always working hard to protect them from risks. The positivists have been attacked for ignoring ethics. Most of the violations on human rights and dignity committed during research studies were linked to the positivists. In a nutshell, according to the social constructionists' worldview on ontology, it is not the majority views and larger sample sizes that should be considered alone but also the individual opinions, values and experiences.

Methodological Worldview of Social Constructionists

The social constructionists make use of inductive approaches when undertaking their projects/studies. They start from the unknown to known and they do not start with theories just like the positivists but they develop theories. The theories in qualitative research are developed under grounded theory design. There are qualitative research methods (derived from social constructionist philosophy) that qualitative specialists make use of. The social constructionists make use of qualitative research designs like the big five traditions i.e. biography, phenomenology, ethnography, grounded theory and case study (Creswell & Clarke, 2007).

In contrast, the positivists make use of quantitative designs when undertaking their studies i.e. experimental designs (quasi and true experiments), non-experimental designs (correlations, survey, ex post facto and descriptive designs). The positivists and the social constructionists differ in terms of their data collection methods and the nature of research instruments. The research instruments of the social constructionists are open ended while the positivists make use of the closed instruments. The social constructionists make use of open ended instruments because they believe that knowledge is difficult to explore and can be only obtained after a detailed/vigorous or intensive fieldwork. The closed instruments of the positivists are not able to get detailed information about the phenomenon because they are predetermined. They do not allow the respondents to think out of the box.

The methods of sampling for social constructionists and positivists vary. The social constructionists make use of non-probability sampling methods that are subjective while the positivists adopt probability methods that are said to be objective. The methods of data analyses are different as well, the social constructionists are concerned with open, axial and selective coding. They are concerned with direct citations quoting the feelings and emotions of individual respondents.

Rhetorical Worldview of Social Constructionists

The social constructionists and positivists differ in terms of their language of reporting. The social constructionists make use of informal language in their reporting. Terms like (I, we, they) and other first person plurals are used in their reports. They make use of the first person plurals because they are supposed to report about the personal feelings of individuals. According to the axiological beliefs of the social constructionists, the individual values, opinions and experiences should be respected. Therefore, the use of the first person plurals in qualitative reports is not surprising.

On the other hand, the positivists are against the use of the first person plurals in research reports because they do not value individual values, feelings, experiences and emotions. They want researchers to use formal language when reporting their findings. In most cases, they say; (*the study observed that.....; it was concluded that .....; the majority of the respondents anonymously agreed that .....*). The authority of the research findings are derived from what the majority views of the respondents and not few individuals. If five respondents are against the view 100 respondents, their view is not considered or taken seriously. The social constructionists dictates that the views of the five respondents out of hundred should be considered and respected. The pragmatist worldview supports anything that works and both formal and informal language may be used provided it works.

Critical Thinking Challenge

1. Which philosophy (social constructionist, positivist or pragmatist) do you support? Why

Tenets of Qualitative Research

A research philosophy shapes and control the way researchers undertake their projects/studies. Qualitative, quantitative and mixed methods studies adopts their characteristics from research philosophies. Qualitative research is a reflection of the social constructionist paradigm. There are several characteristics of qualitative research and readers should appreciate the fact that qualitative projects involves an intensive fieldwork exploring the opinions, emotions and experiences of individuals. This section discusses the key characteristics of qualitative research.

Qualitative research questions are unique in that they are open ended. According to Bryman & Burgess (1993), qualitative research seeks to answer the '*what, how and why*' questions that arise in research. The '*what, how and why*' questions are open ended questions and they allow respondents to express their views in detail about their experiences and emotions. According to the social constructionist paradigm, knowledge and reality in research are difficult and complex to observe. The duty of the qualitative intervention is to give a detailed study about the phenomena that is why their research questions are open ended. Open research questions call for an intensive study compared to the quantified questions.

On the other hand, quantitative research questions are different from qualitative research because they are quantified. The introductory words of quantitative research are '*to what extent, how much, how many, when*'. These questions are quantified because the positivist subscribe to the view that there is knowledge in the universe waiting to be found/observed. The positivists talk of ultimate reality and knowledge that is why their questions are quantified and their worldview of knowledge and reality is simplistic. They believe that knowledge is easy to get contrary to the perspective of the social constructionists.

The major weakness of the quantitative research is that it is not able to answer the 'why, how and what' questions (Madondo, 2016a). A quantitative study can conclude that 16% of women are diagnosed with breast cancer annually but will not be able to give reasons to justify the trend. If a researcher tries to explore the causes, signs and effects of breast cancer, that should be done via qualitative intervention which involves an intensive and vigorous fieldwork. There is need for qualitative and quantitative studies to complement each other. The emergence of mixed methods research is a noble progress in research because it ensures completeness in research studies and brings detailed results.

Qualitative research instrument are open ended. The social constructionist paradigm dictates that there are multiple realities that are difficult to observe and there is therefore need for quantitative instruments to be open ended in order for the researchers to get detailed information about the phenomena under the study. It has been noted within this section that a qualitative study is complex because it is intensive and requires lots of effort in data analysis and gathering.

Quantitative research instruments are closed ended and quantified. Respondents are not given chance to operate outside the box. The social constructionists challenged the positivists for making use of closed instruments that are not able to explore the phenomena in detail. Readers should appreciate the contribution of mixed methods research. Apart from bringing harmony between qualitative and quantitative approaches, mixed methods research ensures completeness in research. The pragmatist paradigm supports anything that works. If combining qualitative and quantitative approaches bring desirable results, mixed methods approaches should be used in research projects/studies.

Qualitative studies are known for their small sample sizes. Small sample size according to Bluman (2007) can be defined as ($n<30$). It is interesting to note that a qualitative study can be credible with a sample of only one respondent. There is a design popular in qualitative study called biography, it involves the study of the experiences of a single person. The aim of qualitative studies is to examine the individual emotions, experiences and feelings. The social constructionists are not concerned about the numbers and majority views. Every individual' views are considered and there is no need to base conclusions only on majority views.

In contrast, the quantitative studies are concerned with larger sample sizes rather than the individuals and minority views. The conclusions made are based on popular views and conclusions are made at the expense of individuals and minority views. There are cases in which majority views are important for progress but individual and minority views should not be ignored. Therefore, the pragmatic approach of undertaking research may be considered. Mixed methods research approaches should come in because both majority and individual/minority views are considered.

Qualitative studies are subjective when dealing with sampling procedures. They make use of sampling procedures that are non-probability in nature. The qualitative studies are criticised for being subjective because the researchers are allowed to interfere with the process of selecting samples from the populations. The social constructionists believe that the researchers cannot be separated from the respondents and knowledge is the product of the interactions between the researcher and the researched in the universe. The quantitative researchers make use of probability sampling procedures. They believe that probability sampling procedures are objective and that there should be no researcher's interference with the researched. However, it is very rare for researchers to be objective on the ground.

The other tenet of a qualitative study is related to the applicability of the research outputs. The qualitative researchers are extremely careful when it comes to the generalizations or application of research outputs to other contexts. The research outputs should only be applied to other similar contexts. The researchers should carefully examine the populations, nature of respondents and methodologies before applying the research findings to other contexts. The post-positivists discourage researchers from generalizing research outputs to other contexts because contexts in research are totally different. There is no need for researchers to generalize the research outputs to other contexts because reality is ever-changing and difficult to get. The quantitative researchers have no problem with applying research outputs to other contexts. They subscribe to the idea of the ultimate reality/knowledge and the generalization of research output is appropriate and normal.

The qualitative researchers use inductive approaches in their studies. They develop theories and concepts that are used to understand the universe properly. They come up with theories via the grounded theory design. The main aim of grounded theory is to develop theories that help the researchers and readers to understand the universe. On the other hand, the quantitative specialists use deductive approaches in their studies. They start from the known from unknown. They test theories rather than developing them. (Hancock, 2000). Readers should appreciate that both inductive and deductive approaches in research studies are appropriate depending with contexts. There is therefore need to consider mixed methods approaches if it works.

The researchers and readers should appreciate that qualitative and quantitative are different paradigms that should be treated with special care. Violations of the rules of undertaking research my result in research output that is not credible. The researchers should be equipped with skills of undertaking ideal qualitative research projects/studies. The purpose of this book is to equip researchers and readers with skills of undertaking ideal qualitative studies. Qualitative studies/projects are unique because they follow their own procedures from the beginning of the study through to the report writing. The disagreements between qualitative and quantitative paradigms should not tempt researchers to disqualify any of the research approaches. Researchers are at liberty to adopt an ideal qualitative/quantitative or mixed methods approach depending on circumstances and contexts. There is no paradigm that is better than the other. The right paradigm should be selected for the right context with the aim of producing desirable research outputs. The emergence of mixed methods research in research as a third force is a positive move but this does not make it becoming a superior approach in scientific research.

Key Areas to Address in Ideal Qualitative Projects

Undertaking an ideal qualitative study can be complicated and confusing. There are key areas that researchers should handle carefully in order for a qualitative study to produce credible and desirable outcomes. The studies related to the qualitative should be done according to the rules of social constructionist approach. This section takes the readers and researchers through key areas that should be considered when researchers are undertaking purely (ideal) qualitative studies.

The Research Questions: research questions are derived from the research objective. They are answered by the literature review and the data collected via fieldwork. A valid qualitative study is the one that answers all the research questions through literature and the data collected from the fieldwork. The nature of the qualitative research questions according to (Hancock, 2000) are the ones which seeks to answer questions on '*why, what and how*'. It is therefore mandatory for researchers to use '*why, what and how*' as leading words to the research questions. The research questions that are quantified should be avoided. The use of the quantified research questions is against the dictates of the social constructionist approach. The following are examples of qualitative research questions;

An example below presents Qualitative Research Questions

- *What is the attitude of university students towards sports?*
- *How is sports connected to academic performance?*
- *Why are university students performing poorly in sports?*
- *What are the strategies put in place to improve the performance of university students in sports?*

Critical Thinking Challenge

1. Select a research problem of your choice and propose four qualitative questions?

The Research Philosophy: research philosophy or paradigm is a key component of qualitative studies. The qualitative procedures of undertaking research are derived from social constructionist philosophy. The specialists for qualitative research should adopt the social constructionist paradigm. Qualitative research is a product of social constructionist paradigm (Creswell,

2014). The qualitative researchers should understand the epistemology, ontology, axiology, methodological and rhetoric characteristics of the social constructionist paradigm. On the other hand, the quantitative and mixed methods researches are linked to positivist and pragmatist philosophies respectively. The qualitative research specialist should select the appropriate paradigm that is connected to qualitative studies.

The Research Designs: the research designs are derived from research philosophies. There are designs that are appropriate for qualitative studies. Creswell & Clarke (2007) suggested five qualitative traditions that are popularly known as the big five. The five qualitative traditions are biography, phenomenology, ethnography, grounded theory and case study. There are other designs for qualitative research but the designs listed above are classified among the key designs of qualitative. It should be also noted that the quantitative designs are experimental (quasi and true experiments) and non-experimental (correlation, survey, ex post facto and descriptive design). The mixed methods according to Kanga, Rutere, Wacera & Burimbui (2015) has four basic designs like embedded, triangulation, exploratory and explanatory. This book concentrated on qualitative designs only because the purpose of the book is to handle qualitative methods and data analysis. The qualitative researchers should only select designs that are recommended by the rules of social constructionist philosophy.

Sampling Procedures: qualitative projects/studies have their own sampling procedures that they use to select samples from the populations. They rely on subjective sampling procedures were all elements are not given equal chances of being selected. The non-probability sampling procedures are linked to the qualitative studies. The qualitative researchers are therefore encouraged to select samples using non-probability sampling procedures in order to fulfil the dictates of the social constructionist approach. In contrast, the quantitative studies rely on probability sampling procedures in which all elements of the population are accorded equal chance of being selected. The probability sampling procedures are objective and they rely on randomization. An ideal (purely) qualitative research project should select samples from the population via non-probability sampling procedures.

Data Collection Methods: there are data collection methods that are appropriate for qualitative studies. In most cases, the following methods are used in qualitative studies; structured and unstructured interviews, focus group discussion, participant and non-participant observation and documentary analysis. On the other hand, the quantitative research's data collection methods are; scales, questionnaires, tests, checklists and others. The qualitative researchers are encouraged to focus on qualitative data collection methods.

The data collection instruments of qualitative research should be open ended. The social constructionist philosophy supports that the qualitative instruments should be open ended. In addition to open ended interview, focus group, observation and documentary analysis instruments, qualitative research can also make use of cameras (for photographing) and tape and video recorders (for recording verbal and visual information) during the data collection. The quantitative research instruments are closed ended. The closed ended scale, tests, questionnaires and checklists are used to collect data. The instruments of quantitative studies are predetermined the respondents have no freedom of sharing their views outside the box.

Data Analysis techniques: qualitative studies have their own ways of data analysis. This textbook takes reader through the steps of qualitative data analysis. The following steps of qualitative data analysis are discussed in detail in this book;

- Data collection instruments
- Data recording and gathering
- Transcription of qualitative data
- Coding (open, axial & selective) of qualitative data
- Report writing

Transcription and coding of data is central to the analysis of qualitative data. The analysis of qualitative data is concerned with coding individual opinions and experiences collectively. On the other hand, the analysis of quantitative data is concerned with use of descriptive statistics as well as parametric and non-parametric statistics. There are software that may be used for qualitative and quantitative data analysis like Nvivo/Atlas and SPSS/ software R respectively.

In a nutshell, qualitative research specialists should consider appropriate methods of handling qualitative studies. They should make use of proper research questions, paradigms, designs, data collection methods, data collection instruments, sampling procedures, data recording instrument, transcriptions and data coding. The qualitative studies should be done according to the rules in order to enhance the acceptable levels of credibility. The researchers and readers will be acquainted with all the information related to the handling of qualitative methods/analyses in this textbook. This will enable them to undertake purely qualitative studies/projects.

Critical Thinking Challenge

Give a brief summary of what an ideal qualitative research look like?

4 QUESTION AND ANSWER SESSION

Is it Important for Researchers to Study Research Philosophy? Some Researchers Believe that Studying Research Philosophy is Useless

It is correct that some researchers are of the view that the studying of the research philosophies is useless. They take such a position because of ignorance (let us not forget that positivist, social constructionist and pragmatist philosophies are linked to quantitative, qualitative and mixed methods research approaches). Qualitative, quantitative and mixed methods research originated from different philosophical positions. A philosophy is an ideology of a research approach/organization/institution/movement. There are a lot of researchers who did not study research philosophies and they are forced to believe that research philosophies are useless because of ignorance. They will only be able to appreciate the importance of research philosophies if they are exposed to them.

The way qualitative researchers undertake their projects is influenced by the social constructionist philosophy. The processes of undertaking research projects vary according to the philosophy adopted. The differences are found on the way the research questions and instruments of a project are structured. The methods and analysis of data are controlled by the philosophies as well. Therefore, reader should appreciate the fact that a research philosophy is the central to any research projects. The researchers sometimes apply the philosophy unconsciously. Researchers and readers need to understand research philosophies in order to appreciate them. Understanding of research philosophies will enable researchers to grasp the methods/approaches of quantitativc, qualitative and mixed methods with minimum challenges.

How Should Researchers Apply Research Philosophies in their Projects?

The application of research philosophies is done at several stages of undertaking research projects. It should be noted that the research questions for projects vary according to philosophies. The social constructionist's research questions are open and not quantified, the philosophy seeks to answer what, why and how research questions. The research questions should be open ended and allow the respondents to express their views and emotions. Ideal qualitative projects seeks to explore (what, why and how) questions in line with the social constructionist worldview. The social constructionist worldview believe in the existence of multiple realities that are difficult to observe, that is why their research questions are open ended.

In contrast, the research questions for quantitative projects are set in line with the positivist philosophy. The research questions are quantified like (to what extent, how much, how far etc.). The positivist philosophy assumes that there is a reality waiting to be observed and the existence of ultimate knowledge. They have a simplistic view of ontology, quantified research questions that are predetermined will enable them to gather data easily.

Mixed Methods Research (MMR) questions combine both qualitative and quantitative research questions. The philosophy behind mixed methods research is said to be pragmatism that believes in the use of anything that works. The quantitative research questions are criticized for not being able to answer the why, how and what questions. The adoption of qualitative research questions can be used to complement quantitative research questions. Mixed methods research is enjoying wide spread usage because of the belief that it harmonizes qualitative and quantitative approaches.

The research designs for quantitative, qualitative and mixed methods approaches vary according to philosophies. The popular social constructionist/ qualitative research designs are biography, phenomenology, ethnography, grounded theory, case study, historical and other designs. These designs call for an intensive study of a phenomenon because of the social constructionist assumption of the existence of multiple realities that are difficult to observe.

The positivist/quantitative designs are experimental designs (quasi and true experiments) and non-experimental designs (correlation, ex post facto, survey and descriptive designs) they are concerned with the quantification of phenomena under the study. The mixed methods designs are triangulation, embedded, explanatory and exploratory. They combine quantitative and qualitative designs. Readers should therefore understand research designs related to each and every research approach. This will enable them to apply research designs in their projects.

The research philosophies can also be applied on research data collection methods and their instruments. The data collection methods and instruments for qualitative projects should be open ended. Qualitative researchers make use of open interviews, focus group discussion, observation and documentary analysis methods. These data collection methods are backed by interview guide, focus group discussion guide, observation guide and document guide that consist of open ended questions. The instruments contain open ended questions because the social constructionist's worldview assumes that the universe contains multiple realities that are difficult to explore and observe.

In contrast, the data collection methods and instruments for quantitative projects have to be closed ended. The data collection methods for quantitative projects are closed ended questionnaires, closed ended checklists, closed ended tests, closed ended scales and other instruments. The instruments are closed because of the belief of the existence of ultimate reality that is waiting to be observed. The pragmatist philosophy believes in anything that works that is why mixed methods research is at liberty of adopting both quantitative and qualitative methods of data collection in their projects.

The application of sampling procedures in research projects is also controlled by the research philosophies. The qualitative sampling procedures are subjective and the process of selecting respondents from the population is controlled by the researcher. Subjectivity in qualitative projects is acceptable, the researcher should interact with the researched (respondents) during the period of sampling and data collection. The qualitative projects should make use of no-probability sampling procedures.

In contrast, the positivist/quantitative researchers make use of objective methods of sampling procedures. The respondents should be selected from the population via randomization. There have to be a distance between the researcher and the researched/respondents. They classify the interaction between the researcher and the researched as a source of bias and subjectivity. The positivist/quantitative researchers make use of probability sampling procedures in their projects. The pragmatist philosophy/mixed method approach has no problem with mixing probability and non-probability sampling procedures. They have to be mixed provided it works and produce desirable results.

Readers should now be able to appreciate the fact that research philosophies govern the way research projects are undertaken. Researchers should be able to apply methods related to an adopted philosophy in their projects. They have to be able to apply methods and approaches basing on the adopted research philosophy. That is why research philosophy is considered to be the center of research projects. The adopted philosophy will enable the researchers to select appropriate methods and techniques for a project.

Which Philosophy (Social Constructionist, Positivist or Pragmatist) is Better than Others?

There is no research philosophy that is better than the other but there may be a research philosophy that is appropriate for a context. There is a temptation among the researchers to declare that pragmatism is the best approach because it is capable of harmonizing the conflict between positivist and social constructionist philosophies. Such a declaration is not always correct because mixed method research still have a lot of gaps that need to be resolved. The declaration that pragmatism is the philosophy behind mixed methods is acceptable to a lot of research scholars but still controversial. Readers should therefore accept the view that all research philosophies are important in research and appropriate philosophy should be selected basing on a context and the nature of a project.

What is the Future of Pragmatism/ Mixed Methods in Research?

The pragmatism philosophy and mixed methods research still have a long way to go and have a lot of gaps. The question on the level of mixing is not yet addressed. Should mixing starts at research questions, designs, data collection methods, sampling procedures, data analysis or both? Some variants of triangulation (like convergence model) talks of equal weighting between quantitative and qualitative but the process of determining equal weighting is not yet clear. In a nutshell, the future of mixed methods is still bleak because of a lot of gaps that are still existing.

REFERENCES

Bluman, A. G. (2007). *Elementary Statistics: A Step by Step Approach.* McGraw Hill.

Borg, W. R., & Gall, M. D. (1989). *Educational Research: An Introduction*. Longman.

Bryman, A., & Burgess, R. (Eds.). (1993). *Analyzing Qualitative Data*. Routledge.

Cherryholmes, C. H. (1992, August–September). Notes on Pragmatism and Scientific Realism. *Educational Researcher*, *21*(6), 13–17. doi:10.3102/0013189X021006013

Cohen, L., & Manion, L. (1994). *Research Methods in Education* (4th ed.). Routledge.

Creswell, J. W. (2014). *Research Designs: Qualitative, Quantitative & Mixed Methods Approaches*. SAGE Publications, Inc.

Creswell, J. W., & Clark, V. L. P. (2007). *Designing and Conducting Mixed Methods Research*. University of Nebraska-Lincoln, Sage Publications.

Crotty, M. (1998). *The Foundations of Social Research: Meaning and Perspective in the Research Process*. Sage.

Farjoun, M., Ansell, C. & Boin, A. (2009). *Perspective—Pragmatism in Organization Studies: Meeting the Challenges of a Dynamic and Complex World*. doi:10.1287/orsc.2015.1016

Guba, E. G. (1990). The Alternative Paradigm Dialog. In E. G. Guba (Ed.), *The Paradigm Dialog* (pp. 17–30). Sage.

Hancock, B. (2002). *Trent Focus for Research and Development in Primary Health Care: An Introduction to Qualitative Research*. Trent Focus Group.

Johnson, R.B., & Onwuegbuzie, A.J. (2004). Mixed Methods Research: A Research Paradigm

Kanga, A., Njeru, L.N., Wachera, E. & Rutere, J. (2015). Rethinking Variant Models of Embedded Research design within a qualitative dominant Mixed Method study. *General Education Journal, 4*(2).

Lincoln, Y. S., & Guba, E. G. (1985). *Naturalistic Inquiry*. Sage Publications.

Madondo, S. M. (2016a). Essentials of Social Science Research. Mount Meru University Publications Unit.

Murphy, J. P. (1990). *Pragmatism: From Peirce to Davidson*. Westview.

Newman, I., & Benz, C. R. (1998). Qualitative-Quantitative Research Methodology: Exploring the Interactive Continuum. Southern Illinois University Press.

Patton, M. Q. (1990). *Qualitative Evaluation and Research Methods* (2nd ed.). Sage.

Phillips, D. C., & Burbules, N. C. (2000). *Post positivism and Educational Research*. Rowman & Littlefield.

Tartakow, D.J. (2012). What is Pragmatic Research? *Ortho Tribune U.S. Edition, 7*(6).

Whose, T. H. C. (n.d.).. . *Educational Researcher*, *33*(7), 14–26.

Chapter 3
Qualitative Research Designs

ABSTRACT

This chapter discusses the big five qualitative traditions/designs and others. It is divided into seven sections, and the first section discusses biography design. The types and steps of undertaking projects via biography design are highlighted in detail. The second and third sections address phenomenology and grounded theory designs respectively. The case study design is exposed in Section 4 while historical design is explored in Section 5. Section 6 contains information related to ethnography design. The readers will be able to understand the data collection and analysis procedures related to each and every design and the unique characteristics of the designs are exposed in this chapter. The chapter is concluded by a question and answer section, where important questions are suggested and answered.

1 KEY CONCEPTS

- **Design** – an overall plan for data collection.
- **Epoche** – the ability of researchers to bracket their feelings and emotions when writing qualitative reports.
- **Epiphanies** – important events in a life of an individual.
- **Primary data** – is the information or data directly provided by the participants or eye witnesses of the events
- **Secondary Data** – is the information/data reported by those who heard from eyewitnesses or the participants.
- **Biographer** – persons undertaking biographic studies.

DOI: 10.4018/978-1-7998-8549-8.ch003

- **Autobiography** – life events of an individual being reported by the same individual.

Learning Outcomes

By the end of the chapter, the readers should be able to;

- Identify data collection methods related to individual designs
- Describe popular qualitative designs
- Connect qualitative data collection methods and related designs
- Distinguish between primary and secondary data in qualitative projects
- Make comparisons between and among designs

Case Study

Before Professor Tom taught his undergraduate students details on research designs, he emphasized the importance of research designs. According to Professor Tom, designs help the researchers to select proper data collection methods/instruments, sampling procedures and the right procedures of processing data. He concluded by saying that the identification of research designs will enable researchers to identify methods related to them. By the end of reading this chapter, readers should be able to understand methods/strategies connected to different qualitative designs.

2 INTRODUCTION

I agree with Prof Tom. An understanding research designs enables students to know the methods related to them. This will make the process of undertaking qualitative projects to become much easier. The readers should now be able to appreciate the contribution of social constructionist, positivist and pragmatist philosophies in shaping the ways research projects are being undertaken. These philosophies gave births to several designs that are currently shaping the conduct of research. Positivist philosophy is supported by experimental designs (quasi & true experiments) and non-experimental designs (correlation, survey, ex post facto and descriptive designs). These are the popular designs for quantitative research. Mixed methods research (a product of pragmatist philosophy) has four key designs like embedded, triangulation, explanatory and exploratory.

Research designs are important because they help the researchers to select proper data collection and analysis methods. Research design is an overall approach for data collection and designs control and shape the processes of undertaking research projects/studies. This section discusses key designs for qualitative research. The readers will be able to understand the key qualitative traditions which are popularly known as the BIG FIVE and other designs. This section is divided into the following sections;

- The Biography Design
- The Phenomenology Design
- The Grounded Theory Design
- The Case Study Design
- The Historical Design
- The Ethnography Design
- Question and Answer Session

The researchers and readers will be able to understand the key tenets and processes of handling each and every design. They should be also able to master the data collection methods appropriate for each design as well as the proper methods of analyses of the data from the fieldwork. Qualitative research designs are selected depending on the nature of the study, environment of the study and the philosophical positions that the researchers adopt. If projects are undertaken in ethnic or race contexts, the researchers may have to make use of ethnography design. In some cases, the researchers may want to undertake ideal qualitative projects and they are forced to select and make use of qualitative methods. Some projects/studies with the aims of exploring what happened in history may adopt historical design.

The Biography Design

Out of all the qualitative designs, biography is unique because it is the only design that undertakes research with only one subject/participant or respondent. It has already been noted that qualitative studies unlike quantitative studies are not concerned with bigger samples but the feelings and experiences of individuals. According to Denzin (1989), biography is the study of an individual's life story and the study may be undertaken by another person or the same person. The researchers who specialize in biography writing are called biographers. The study of biography involves a reconstruction of an individual's life events.

There are several types of biography according to Cole (1924) and they differ in terms of their purpose and procedures of undertaking them. Autobiography is among the types of biography. The term autobiography was first coined by William Taylor in 1797 and the origin of autobiography can be traced to the 1790s. The life accounts are written by individuals themselves. Biography deals with an individual's outstanding life experiences.

However, autobiographies have been criticised for their lack of objectivity. They fail to balance between the positive and negative sides of the stories. It is difficult and rare for individuals to write objectively about their dark past and life of controversies. Even though autobiographies have been questioned on objectivity, they provide genuine accounts of individuals' lives because the individuals will find it easy to record and account for their life experiences.

The other type of biography is known as biographical studies. Unlike the autobiography, biographical studies narrate an individual's life story as seen by another person. It is not the same individual who is involved in the tasks of data collection and report writing but another person should be responsible for undertaking such kind of a research project. There are several sub types of biographical studies like popular, fictional, literary and reference biographies. Biographical studies are popular with university students.

The biographical studies are claimed to be more objective compared to the autobiography. They are able to handle and suppress biased feelings of an individual involved. The biographers are able to give a balanced narrative of an individual's life from a neutral perspective. The biographers will be able to triangulate data collection methods in order to provide a balanced account of an individual's life events.

Memoir is another popular type of biography. It is sometimes called life history. It is an account of an individual's life and its link on important historical events or phenomena. It narrates the popular events or situations that took place during the time that an individual survived. There are several types of biographies but autobiography, biographical studies and memoir are among the popular ones. The biographers should adopt an appropriate type of biography when planning for research projects related to the nature of this design.

There are several stages that biographers should follow when undertaking a biographic studies. The stages discussed in this sections are a modification of Denzin (1989)'s proposed steps of undertaking biographical projects. A biography study starts with the setting of objectives. The objectives should be chronologically sorted i.e. exploring an individual's life from education, family background, employment history, activism and death. The objectives should be set following a chronological pattern. The purpose of the biographical study is to give answers to the objectives set and biographers should gather materials related to the objectives set.

The second step will be to gather epiphanies or key events that are related to the objectives set. This is a stage of data gathering. There are several methods that may be used to gather data like face to face interviews, video recordings, photographing, documents (diaries, emails, letters and newspapers). The persons close to an individual participating in a biography project may be interviewed as well. The biographers should make sure that key informants are identified. The friends, spouses, children, workmates and siblings may be part of key informants. There is need for biographers to triangulate several methods in order to provide a balanced view of what actually happened in an individual's history.

After the gathering of data, the biographers should start a process of data analysis. The categories should be formed based on epiphanies. Only epiphanies should be categorised and coded for the purpose of report writing. An open coding of epiphanies should be done following the rules. However, the open coding should be done after transcription process has been completed. The biographers should make use of proper transcription methods. The processes of transcriptions are discussed in detail in this book.

The analysis of data should be followed by the report writing. The key findings and conclusions should be based on the study objectives and the key epiphanies. The biographers should bracket their own feelings and the voice of an individual under the study should prevail. However, the biographer should know when and how to interfere with the study. According to the rules of qualitative research, researchers and the researched are inseparable and the interference of the researcher with the study is acceptable but should not compromise the results.

Biography design is popular among the qualitative traditions. Biographers should acquire skills on handling biographic studies. Appropriate biography types, data collection and analysis methods should be used. The steps of handling biography designs should be followed and biographers are encouraged to triangulate methods in order to provide balanced accounts of individuals' lives in history.

Critical Thinking Challenge

1. Describe steps of undertaking a biographic project in qualitative paradigm?
2. Why should readers trust the outcomes of a project done by a single respondent?

The Phenomenology Design

Biography design is concerned with the narrative of an individual's life experiences while phenomenology explores the lived experiences of a number of individuals. The term phenomenology was derived from the discipline of philosophy. Phenomenology as a philosophy discipline originated from Edmund Husserl (1859-1938). He was born on the 8th of April 1859 and died on 27th of April 1938. He studied at university of Vienna and obtained a doctoral degree in mathematics in 1882.

Generally, phenomenology is the study of phenomena or phenomenon. The main aim of phenomenology is to describe experiences as they were lived. In research designs, phenomenology is a way of interpreting events and their relationships as they were lived by people. There are a lot of events in our societies today which are not fully explained and these events should be fully interpreted through phenomenology. Phenomenology helps people to be aware of their environment and world.

The year 1901 was important in the history of qualitative research; it was the year that Edmund Husserl used a method known as phenomenology in his logical investigations but the method was fully developed in 1913. Phenomenology helps people to know their world much better. The methods of data collection for phenomenology designs are mainly interviews.

There are key issues that the researchers making use of phenomenological design should know in order to undertake successful projects of phenomenology. According to Field & Morse (1985), the researchers undertaking a phenomenological study should practice epoche. Epoche means the bracketing of the researcher's feelings, emotions and experiences so that they cannot interfere with the study findings and conclusions. The researchers may be tempted to expose their lived experiences on the phenomena under the study at the expense of the research subjects' lived experiences. Such kinds of interferences may only be dealt with through the practising of epoche by the researchers.

Another area to consider in phenomenological studies/projects is the selection of key informants. A phenomenological study is an intensive fieldwork that gathers data from the key informants. Key informants are the people directly affected or had an encounter with the phenomena. The sample size for phenomenology study may consist of 5 to 25 respondents (Polkinghorne, 1989). However, other authorities believe that the maximum sample size may be not be more than 10 respondents. The qualitative designs are concerned with the feelings and experiences of individuals and not the quantity of respondents. The researchers may identify gatekeepers who can help them to identify key informants. The process of identifying key informants may be hectic. The selection of wrong informants may affect the credibility of the study.

A key method of data collection (for phenomenological projects) is an interview. The interviews commonly used are structured. The interview session should be designed in such a way that the lived experiences of individuals about the phenomenon are captured with minimum interferences from the interviewers. Interrupts are likely to disrupt the thought processes of respondents during the interview sessions. The interviewers should give the researchers enough time to share their lived experiences with no or minimum interruptions during the interview sessions.

The example below is an illustration of how the interviewers should handle interviews when gathering data for phenomenology studies;

- *Interviewer: my name is James, undertaking a study on the impact of famine that took place in 2009. I know that you were among the victims of 2009 famine. May you please share with me your experiences of the 2009 famine? I will give you 20 minutes to respond and you are free to ask questions anytime.*
- *Interviewee: I will never forget the famine of 2009, I lost my relatives during that sad period in my life ....*

There are lessons that researchers undertaking interviews for phenomenology studies should know when undertaking their interview sessions. The interviewer should avoid interruptions when the interviewee is speaking. This is done in order to avoid the interruption of the flow of the lived experiences. The interviewee is given 20 minutes to speak and requested to ask questions that need clarification anytime. However, the interviewer may have a minimum interference during the interview session in order to keep the interviewee focused but the interference should not disturb the flow of thoughts of the interviewee.

Undertaking a phenomenological study may be complicated if the researchers are not knowledgeable with phenomenology from the philosophical positions. The researchers are therefore requested to read a lot on phenomenology before undertaking their studies. The design originated from philosophy and researchers need to acquire enough knowledge before progressing to the next level.

Another challenge is on practising epoche. It is a very difficult task for researchers to bracket their feelings about the phenomena under the study. Personal feelings and emotions of researchers are always difficult to control. There are no clear procedures laid down on how researchers should practice epoche. Bracketing of researchers' feelings appear to be more psychological than being practical.

The other challenge is on the methods of administering interviews for phenomenology. The researchers may disturb thought process of an interviewee during interview session if there are interruptions. The problem comes if the researcher is not skilled in handling interviews for phenomenology. The interviews are complicated to administer.

The data gathered from interviews should be transcribed and categorized based on the lived experiences, statements and meanings. The open coding should be done basing on the lived experiences and the conclusions and findings should be linked to the objectives of the study.

Critical Thinking Challenge

1. The interviews for phenomenology are unique when compared to interviews for other qualitative designs. Describe the uniqueness of phenomenology interviews and briefly give insights on how they are conducted?

The Grounded Theory

The main aim of grounded theory is to develop theories. The role of biography is to narrate the life experiences of an individual. Phenomenology explores the lived experiences of a group of individuals. The role of grounded theory is to formulate theories. In quantitative research, theories are not developed but they are tested. Gleser and Strauss (1967) were the brains behind the grounded theory. Strauss and Corbin (1990) made enormous contributions on grounded theory. They polished the grounded theory and gave it its fuller picture. Grounded theory originated from the discipline of psychology.

Various methods can be used for data collection like interviews and observation. Data is collected and analysed through a process known as constant comparative (Creswell, 1998). The proponents of the grounded theory believe that theories are grounded in the data. It is therefore the duty of researchers to expose and develop such theories from the data through a process known as constant comparative.

Constant comparative is the repeated collection and immediate analysis of data on several occasions till saturation is achieved. Grounded theory is different from other designs because, data is collected and analysed immediately and this is done in order to identify emerging issues. After the identification of emerging issues, the researcher has to go back to the field to collect data till saturation is achieved. The researchers should go back to the field to carryout interviews for 20 to 30 times till saturation is achieved (Creswell & Clarke, 2011).

There is need to define the term saturation in grounded theory. Saturation is a stage where there are no longer emerging issues arising after successive data gathering and data analysis, that is the point one can confidently say, a theory has been developed. If emerging issues are constantly appearing soon after data analysis, saturation will not have been achieved. Saturation is reached after continuous and data gathering and immediate analysis. If there are emerging issues arising it means saturation has not been reached. The point of saturation is the point where theories are developed (Boeije, 2010).

The grounded theory is unique in terms of the way the field data is analysed. The data analysis is done in three processes of coding. The first phase of coding is called open coding. It is the initial coding in grounded theory were the gathered data is classified into categories. The next phase of coding under grounded theory is called axial coding. Axial coding is the re-categorization of data after the initial coding. The open codes are further classified under new codes or categories. The last level of data analysis under grounded theory is called the selective coding. Selective coding is the selection of a key category that can connects all categories. The steps of open, axial and selective coding are highlighted in chapter 7 of this textbook.

Open, axial and selective coding were initially developed to analyse data for studies adopting grounded theory design. However, open, axial and selective coding can be performed outside the grounded theory. The researchers are encouraged to perform the three coding outside the grounded theory though the three processes of coding were initially design for the grounded theory.

The Historical Design

According to Ronald (2007), the purpose of historical designs is to collect, verify and synthesize evidence from the past to establish facts that defend or refute hypotheses. Generally historical design is a product of historical studies and the aim is to be able to understand what happened in the past. Scholars cannot close the doors of history because the events are past, studies in history can be done in any field of study. These are studies based on what happened in the past in social sciences and those studies should be supported by historical design.

Sources of Data

There are two major sources of data of historical design;

- Primary data
- Secondary data

Primary Data

These are the events directly reported by the eye witnesses or the observers or the participants. Researchers trust historical data on condition that the data was recorded by the eye witnesses, participants or observers. The historians are of the view that their data is genuine and live because the sources witnessed the events directly, they observed what was happening and they sometimes participated on these events.

Even if the data was reported by the participants or eye witnesses, queries are raised on the correctness of the data. The events took place in the past and sometimes the eye witnesses may find it difficult to remember these events. Even if the data is kept as documents, the safety of such documents may be compromised as time progresses that is why readers are always questioning the credibility of the historical data. The other problem is that, the eye witnesses or participants may be overwhelmed by personal and emotional biases.

It is difficult to control these past events because they have already taken place. What makes science empirical is the ability of the researchers to control the events but because historical events are past, it may be very difficult to control the events.

Information for primary data is obtained from variety of data sources. There are several ways in which primary data can be gathered. The data can be gathered from documents. These are past written records documented by the observers, participants or the eye witnesses of these events. For example the constitutions, court decisions, official minutes, autobiographies, letters, diaries, personal will, affidavits, certificates, bills, receipts maps videos, pictures, paintings, research reports. Most of these documents are kept in archives, libraries, museums and other places. They are always available for researchers and can be accessed freely or sometimes at low costs.

The documents contain the information about various events useful for research. The researchers should make use of these documents when the need arises. In some institutions researchers are not allowed to quote scholars or information that are more than ten years old. But in historical researches these outdated documents and information are useful and relevant. Oral testimonies are also part of primary data sources. Oral testimonies or interviews given to the eye witnesses are recorded and kept as videos, cassettes, tapes etc. They provide live accounts and genuine information of the past events. The information is useful to true historians and is available and accessible for academic use. Part of the information may be found on internet.

Remains of the past are also reliable sources of primary data. The remains of the past are sometimes called relics and students of archaeology depend very much on relics. In each and every period people use different clothing materials, coins, building materials, furniture, photographs, weapons and those materials are helpful for historical studies. One can be able to make a conclusion on what was happening in the past through studying them.

The relics are kept safely in museums and students of research may make use of them in their studies. The study of these remains is a clear indication that historical events are empirical because researchers can still see, touch, feel and test these remains in laboratories. A genuine research should be based on empirical evidence.

Secondary Data

Unlike primary data, secondary data/events are not witnessed by the reporter. The reporter was not on the scene and is based on what the historians heard from people who were either observers or participants/eye witnesses. The reporters may be informed by people who got information from participants/observers. The textbooks, letters and diaries can be primary or secondary data source depending on whether they were written by a participant/observer. If the book is not written by an observer/eyewitness, therefore, it can be classified under secondary data. Even the oral testimonies can be secondary as well if they were testimonies given by people who were not observers/participants. Oral testimonies can be primary data source if they were given by the participants/observers.

Knowing the appropriate research designs assist the researchers to select proper data gathering and analysis tools. Researchers adopting historical designs should make use of primary and secondary data sources in their data gathering. It is always difficult for researchers to make use of interviews and questionnaires when handling historical studies. The transfer and extinction of subjects may be a challenge.

Critical Thinking Challenge

1. What is your assessment of primary and secondary data in qualitative projects?
2. What are the challenges of using interviews on qualitative projects that took place more than 20 years ago?

The Case Study Design

A case study involves an extensive study of a bounded system. It is bounded by place and time. The design originated from sociology and anthropology disciplines. According to Hancock (2002), just like survey, case study can be used in both qualitative and quantitative studies. It can examine an institution or a single person. Case study is unique in the sense that it is the only design of qualitative that is supposed to be accomplished by the use of several data gathering methods.

The first step for qualitative researchers to accomplish is to identify the type of a case study to be adopted by a project. A case study may involve a study of several programs and such type of a case study is called multi-site study. On the other hand, it may examine a single program and the type of such kind of a case study is called within-site study. A multiple case study and single case may be used in place of multi-site study and within-site study respectively. The duty of the researchers is to check on whether a study involves a single or multiple cases. This will help them to strategies on proper methods that can be used to execute the task.

Stake (1995) talks of intrinsic case study and instrumental case study. An intrinsic case is a case studied because of its uniqueness. For example a study of a TanzaniteOne, the only mining company of Tanzanite in the world. There is no any place in the world where Tanzanite mineral is found. Stake (1995) used the term collective case study to refer to the study of multiple cases. Collective case study is not different from multi-site study or multiple case study.

Hancock (2002) came up with another type of a case study called extended case study. Extended case study is the study of case or cases/event or events over an extended period of time. The key characteristic of the extended case study is that the study is undertaken from the beginning to the end with the same respondents. It is very important for researchers to understand the nature of a case study to be adopted. The methods of data collection and sampling techniques to be used vary according to the type of a case study. Multiple cases use different methods of undertaking studies as compared to single case studies.

There are two critical issues that the researchers of case studies should know before undertaking their projects. The two issues are critical because they determine the credibility and the trustworthiness of a qualitative study. The first critical issue to consider is on sampling techniques. The researchers should make use of non-probability sampling procedures. The type of sampling procedure recommended is purposive sampling. There are several types of purposive sampling discussed in this book and researchers are encouraged to select appropriate sampling procedures for purposive.

In case of studying a multiple case study, researchers should opt for purposive sampling procedures that represent the entire bounded system like heterogeneous sampling. In case of studying a single case, heterogeneous sampling may not be necessary. Readers should appreciate at this level the importance of selecting proper sampling procedure as well as the essence of identifying the nature or type of a case study. The sampling procedures for single and multiple case studies are sometimes different. Special attention should therefore be given to the nature of sampling procedures to be adopted.

The second issue to consider about the case study is the types of data collection methods to be used. It has already been noted that case studies are unique in that they use several data collection methods. According to Yin (1994), a case study may use as many as six data collection methods. He suggested participant observation, direct observation, interviews, document analysis, archival records and physical artefacts. Other methods of collecting data like focus group discussion may be used as well. A case study involves an intensive study within the research site and several data collection methods are used in order to understand the case under the study. For a researcher to make use of only two methods of collecting data under a case study may not be enough. The credibility of such a study may be questioned because case studies are intensive in everything.

The data analysis of a case study can be holistic or embedded. The analysis should be based on the entire case if it is holistic and embedded analysis is based on specific cases within a case study. The processes of open coding should be based on specific cases within a case or may be holistic. The analysis techniques should be determined by the nature or the type of a case study, which is why researchers are encouraged to be cautious when handling the data collection methods, sampling techniques and the data analysis methods.

The researchers should also understand the challenges of handling case studies because the challenges have a serious impact on the trustworthiness of their studies/projects. From the social constructionist perspective, the results of case studies should not be generalized because cases are totally different. The positivist philosophy (quantitative) has no problem with the generalizations of research results. However, generalizations of research findings according to qualitative practise should be done only to similar contexts. In case of dealing with cases, generalization of results (for cases to other contexts) is strongly prohibited because contexts are totally different.

The challenge of sampling is inevitable if the researchers are not skilled in handling case studies. The use of different types of cases studies is correlated to different types of non-probability sampling procedures. There is therefore need for qualitative researchers to get enough skills on the handling of sampling techniques in case studies. The same applies to the selection of data collection methods. The researchers should make sure that appropriate and adequate data collection methods are selected. The trustworthiness of a qualitative study is determined by the data collection and sampling methods. There are also problems related to the analysis of cases within a multiple case study and justice should be done in order for the analysis to be a reflection of the entire case. The qualitative researchers should not therefore take a case study approach for granted.

Critical Thinking Challenge

1. Case study design is said to be unique. In what ways?
2. Should researchers generalise (apply to other contexts) qualitative outputs from case study design? Justify

Ethnography Design

The anthropologists are the pioneers of ethnographic studies. The main aim of ethnographic design is to describe and examine cultural characteristics. Ethnography involves studies related to culture, ethnicity, tribes, religions and regions. According to Hancock (2002), people under examination should have something in common based on geographical location, tribe, religion and shared experiences. Studies related to ethnography are also connected to race and ethnicity.

Ethnographic studies are difficult to apply to different contexts because culture varies from one place to another. Qualitative research findings are very difficult to generalize just like the findings from ethnographic studies because of the variation of culture from one community to another. It is in this context that qualitative studies are reluctant to allow researchers to apply study findings to other contexts. Application of ethnographic findings to other contexts may be wrong

According to the guidance from Creswell (1998), culture consist of behaviour, language and artefact. The studies of ethnography revolve around the behaviour of people, the language used by those people as well as the materials that they use. Just like case studies, ethnographic studies are intensive and several people are supposed to be interviewed.

There are several issues that the researchers handling ethnographic studies should consider before undertaking ethnographic studies. The first issue to consider is related to reactivity effect. The people on the research site may be hostile to the researchers visiting their grounds. In some cases, injuries or even deaths may be reported. The researchers should observe some ethical principles in order to avoid inconveniences.

To deal with the challenge of reactivity effect, the researchers are supposed to identify gatekeepers. Gatekeepers are the people in charge of the research sites. They are important in that they can shield the researchers from harm during the period of data collection or the fieldwork. The gatekeepers are supposed to introduce them to the research subjects on the research sites.

The gatekeepers are respected people in the research sites and they are there to maintain peace between the researchers and the researched.

The selection of gatekeepers is followed by the selection of key informants. Key informants are the right people in the research sites who can provide the appropriate data during the fieldwork. The selection of wrong respondents has an impact on the credibility of the study. The gatekeepers are important at this phase because they can help the ethnographers to identify the key informants. The gatekeepers are well versed with what is happening on the research sites and the ethnographers should take advantage of this.

The other issue to consider for ethnographers is the idea of reciprocity. Reciprocity means the symbiotic relationship between the ethnographers and the people in the research sites. The ethnographers/researchers will get data from the people on the research sites and they should also reciprocate or give back. In some cases, the ethnographers give some small gifts like cigarettes or food but they should not be spoiled. Spoiling them (research subjects) may spoil the credibility of the study.

There are several methods of data collection that may be used when undertaking an ethnographic study. Participant observation is claimed to be the best and a suitable method to employ on studies connected to ethnography. The researchers/ethnographers undertaking such studies should be from the same culture or embedded in a group. This will help them to interact with the people on the research site smoothly. Such ethnographers are well versed in the local culture (behaviour, language and lifestyle). A lot of information related to participant observation is discussed in the data collection methods chapter.

There are several challenges that are associated with ethnographic studies. The researchers should be well grounded in anthropology. It is therefore vital for researchers to get some skills related to anthropology. The other challenge may be related to reactivity. The researchers should follow ethics in order to deal with the challenges related to reactivity. Researchers should be well versed in the culture of the people within the research sites and should avoid provocation of local culture and their people. Acceptability of researcher is important when undertaking ethnographic studies.

Readers are encouraged to read chapter one of this book. The chapter gives practical experiences of ethnographers with subjects on the research sites. It exposes how the ethnographers dealt with reactivity, reciprocity and the selection of gatekeepers and respondents during data gathering.

3 QUESTION AND ANSWER SESSION

Why Should Qualitative Researchers Understand Designs?

Qualitative and quantitative designs are different and the difference is based on the philosophical position they adopt. The understanding of qualitative designs will enable researchers to pick appropriate methods for data collection methods/instruments, sampling techniques and analysis procedures. There are appropriate data collection methods for qualitative designs as indicated in chapter four. It should be known that a case study design has several methods of data collection because it involves an intensive study of a bounded system. Document analysis is a key method of data gathering for a historical design.

For those undertaking projects with grounded theory, they should be able to perform open, axial and selective coding. The analysis of data gathered under biography design should be done basing on identified epiphanies. When readers are studying qualitative designs, they should be able to know the data collection methods, sampling procedures and data analysis techniques attached to them. A detailed study of qualitative designs will enable researchers to undertake their projects with minimum challenges.

What is the Uniqueness of Each and Every Qualitative Designs?

Each and every qualitative design is unique. Biography is the only design that undertakes a project with one respondent. Qualitative designs are not concerned with the numbers but with the experiences and emotions of few individuals. That is why a design like biography can study the life experiences of an individual. Phenomenology design is unique because it studies the lived experiences of a group of few individuals. The researchers undertaking projects with phenomenology should acquire skills of handling phenomenological designs as they are implemented in philosophy.

Grounded theory is unique in three ways. First, it is the only design designed to generate theories. Second, it is the only design that is undertaken via the constant comparative strategy. It involves a multiple collection of data and immediate analysis till saturation is reached. Finally, grounded theory data is analysed via open, axial and selective coding. Qualitative researchers are encouraged to study grounded theory in order to grasp the practical skills of open, axial and selective coding.

A case study design is unique in that, it is the only design that make use of several data collection methods. A case study project may use above five data collection methods. It is intensive in terms of data gathering that is why several data gathering methods are used. Ethnography design is unique and appropriate for studying ethnics, tribes or a group of people with shared experiences. It is therefore important and mandatory for qualitative researchers to study research designs in detail because they assist them to select appropriate methods for their project.

Is a Project Undertaken with one Person Credible?

It is correct that some qualitative projects can be undertaken with a single respondent. The practice is common under biography design and it studies the exceptional life of an individual. Undertaking a project with a single respondent is acceptable under qualitative research. The results of a biography design are credible and qualitative studies are not concerned with numbers but emotions and experiences of individual respondents. Readers should be able to appreciate that qualitative projects are undertaken with few respondents. The concern of the qualitative researchers is not to make decisions and conclusions basing on majority view but to examine the emotions, feelings and experiences of few individuals. The sample size for quantitative projects is high and can study even 1000 respondents. A sample size of less than 30 respondents is considered unfit for a quantitative project. However, a qualitative study can undertake a credible study with a single respondent.

Why are Interviews for Phenomenology Design Unique?

An interview administration of phenomenology projects is very different from the way interviews are administered under other qualitative designs. The aim of the phenomenology design's interviews is to tap the lived experiences of few individuals without disturbing their thought processes. The interviewer should not interrupt when the interviewee is sharing his/her experiences. Frequent interruption may disturb the thought process of an interviewee. The interviewer is supposed to introduce him/herself, pose a question and give the interviewee enough time to respond without or with minimum interruptions. The interviewers should therefore be careful when administering interviews for phenomenology projects. They have to acquire skills of handling such interviews.

How did Coding and Grounded Theory Originate in Qualitative Research?

There is a popular view among the qualitative researchers that open, axial and selective coding in qualitative research originated with grounded theory and the development of the constant comparative concept. If we are to follow the work of Glaser & Strauss of 1967, the proposal that coding originated in 1960s may be making sense to many qualitative researchers. The development of grounded theory is linked to Glaser and Strauss that is why many qualitative researcher believe that coding in qualitative research originated with the development of the grounded theory in 1960s.

However, the author of this book is of a different opinion. The author is of the belief that coding in qualitative research started in the discipline of sociology in 1930s. It was a period when the analytic induction concept was developed by sociologists. This question and answer session seeks to demonstrate that constant comparative analysis in grounded theory was borrowed from the analytic induction process.

What is analytic induction? It is a concept that originated from sociology. The proponents of analytic induction are Znanieki (1934) and Cressey (1953). Znanieki proposed the concept of analytic induction and Cressey gave it flesh. Analytic induction is defined as the systematic and exhaustive examination of few number of categories. The key word here is exhaustive and in grounded theory's constant comparative they use the word saturation in place of exhaustive. When taking about exhaustive or saturation, the implication is that the data should be examined/analysed till emerging issues are not coming up.

There are several stages of analytic induction but they are not limited to the following steps (suggested in this paragraph). The initial stage of analytic induction is the identification of a phenomenon or a problem that needs to be addressed. The identification of a phenomenon is followed by the development of hypotheses. The research team should undertake a fieldwork to test the hypotheses. The initial gathering of data to test hypotheses should begin. The immediate analysis of data may provoke the research team to formulate new hypotheses depending with the emerging issues. The process may go on and on till an exhaustive examination of the data. In constant comparative and grounded theory, the examination should continue till saturation is reached.

Why is Analytic Inductive Important in Qualitative Data Analysis and Coding?

It led to the development of the processes of open, axial and selective coding in grounded theory. The concept provoked the development of grounded theory and is similar to the constant comparative concept of grounded theory. The analytic induction calls for multiple data collection leading to exhaustive examination just like the concept of constant comparative of grounded theory. It calls for the examination and analysis of data till saturation is reached. The processes of undertaking constant comparative are closely related to the processes of undertaking analytic induction.

What is Constant Comparative in Grounded Theory and Qualitative Data Analysis?

Constant comparative is a concept of grounded theory designed to analyse data till theories are exposed. It is a continuous data gathering and immediate analysis till saturation is reached. In other words, constant comparative is the exhaustive examination and analysis of the gathered data till a theory is developed. The duty of qualitative researchers is to expose theories via constant comparative procedures.

What are the Steps of Constant Comparative in Qualitative Data Analysis?

There are several steps of data analysis of qualitative data basing on constant comparative. This section discusses four steps of constant comparative data analysis according to Boeije (2010). She called the first step exploration. It involves the identification of codes and the researcher should discover concepts linked to the data. The second step is defined as specification stage. It involves the selection and development of key concepts and the similarities and differences among concepts are observed. The third stage is known as reduction. This is a stage in which a core concept is determined. The core concept will be compared against all other concepts. The final stage is the development of a theory and Boeije (2010) call that stage integration.

A close look on the constant comparative stages reflects open, axial and selective coding of qualitative data. The exploration stage is related to open coding which is the identification of several and initial categories. The specification stage is connected with the axial coding in which the initial categories are re-categorized. It involves the second coding of initial categories. The reduction stage is linked to selective coding. It is the stage in which a core category is identified from the categories. The core category should then be compared with other categories.

There are striking similarities between constant comparative and analytic induction strategy. They both call for an exhaustive examination and analysis of data as well as multiple data gatherings and immediate analysis of the data. There is no doubt that analytic induction provoked the development of constant comparative if we are to consider the dates of their origin. The aim of constant comparative in data analysis is to analyse the data exhaustively. The same should happen to the data analysis of other qualitative designs. Researchers are encouraged to use coding (axial, open, selective) outside the grounded theory as well. Details of constant comparative and analytic induction are found in chapter seven of this book.

REFERENCES

Boeije, H. (2010). *Analysis in Qualitative Research.* Sage Publications Ltd.

Creswell, J. W. (1998). *Qualitative Inquiry and Research Design: Choosing among Five Traditions.* Sage.

Creswell, J. W. (2007). *Research Design. Qualitative, Quantitative and Mixed Methods Approach* (2nd ed.). Sage Publication.

Creswell, J. W., & Plano Clark, V. L. (2011). *Designing and Conducting Mixed Methods Research* (2nd ed.). Sage.

Denzin, C. D. (1989). *Interpretive Interactionism.* Sage.

Glaser, B. G., & Strauss, A. (1967). *Discovery of Grounded Theory.* Aldine.

Hancock, B. (2002). *Trent Focus for Research and Development in Primary Health Care: An Introduction to Qualitative Research.* Trent Focus Group.

Marshall, C., & Rossman, G. B. (2010). *Designing Qualitative Research* (5th ed.). Sage Publication.

Merriam, S. B. (1998). *Qualitative Research and Case Study Applications in Education*. Academic Press.

Polkinghorne, D. E. (1989). Phenomenological Research Methods. In *Existential-Phenomenological Perspectives in Psychology* (pp. 41–60). Plenum.

Stake, R. E. (1995). *The art of case study research*. Sage.

Strauss, A., & Corbin, J. (1990). *Basics of Qualitative Research: Grounded Theory Procedures and Techniques*. Sage.

Yin, R. K. (1994). *Case study research: Design and methods* (4th ed.). Sage.

Chapter 4
Qualitative Data Gathering Instruments and Methods

ABSTRACT

The chapter is divided into five sections. The first section discusses the interview method from the social constructionist and qualitative perspectives. The second section contains information related to the procedures of undertaking focus group discussion in qualitative projects. The different types of observation methods are highlighted in Section 3, but only participant and non-participant observation variants are discussed. The primary and secondary documents are addressed in Section 4, and examples are given. The chapter is concluded by a question and answer section, where important questions are raised and answered in a simple way.

1 KEY CONCEPTS

- **Observation** – a systematic examination of events as they happen on the ground by the observer.
- **Observer** – person examining events as they happen on the ground.
- **Interviewer** – a person moderating an interview session.
- **Interviewee** – person participating in an interview session.
- **Group Discussion** – interview session with two or more participants.
- **Bio Data** – information or data related to demographic characteristics.
- **Transcription** – the compilation or writing down of data recoded from the research sites.

DOI: 10.4018/978-1-7998-8549-8.ch004

Learning Outcomes

By the end of this chapter, readers should be able to;

- Describe popular data collection tools for qualitative projects.
- Describe and justify ethics involved in data gathering of qualitative projects.
- Understand characteristics of qualitative instruments.
- Appreciate steps involved in gathering qualitative data.

Case Study

Grey, a fourth year student at The University of Zambia wrote a research proposal on the challenges facing female single parents. The proposal was successfully approved by the supervisors. During a focus group discussion session with six participants, Grey asked one participant to shade light on her private affairs. The woman overreacted and pulled out of the group interview. The group interview session was eventually abandoned because three other respondents decided to pull out in solidarity.

2 INTRODUCTION

Grey's experience during the focus group discussion should remind readers that administering group interviews may be a challenge if the interviewers are not skilled. There is therefore need for researchers to acquire proper skills for data gathering in order to avoid disappointments. The nature and structure of qualitative and quantitative research instruments and methods are totally different. Quantitative studies make use of closed ended instruments like scales, checklists, questionnaires and tests when gathering their data. The quantitative instruments are closed ended. The positivists believe in the assumption of ultimate knowledge and reality waiting to be observed. The instruments are closed ended because of their belief that knowledge is predetermined. On the other hand, the social constructionists (qualitative specialists) assume that knowledge and reality are fragile and difficult to obtain. They support the use of open ended instruments because knowledge is difficult to gather. This is the reason why qualitative researchers support the use of open ended research instruments.

There are several data gathering methods for qualitative research projects. Key among the data gathering methods are interviews, focus group discussion, observation and document analysis. These methods have their instruments that are used to collect data from the research sites. The following can be used as instruments of qualitative projects; observation guide, focus group discussion guide, interview guide, audio/video recorders and cameras.

The main objective of this chapter is to enable readers and researchers acquire practical skills of managing qualitative data gathering processes. Therefore, this section is divided into following sections;

- Interview Method in Qualitative Research
- Focus Group Discussion Method in Qualitative Research
- Observation Method in Qualitative Research
- Documentary Analysis in Qualitative Research
- Question and Answer Session

Interview Method in Qualitative Research

There are several ways in which interviews are administered to research subjects. The interviews may be administered via telephone. Such kinds of interviews are called telephone interviews. Other interviews are undertaken via technological gadgets like laptops or desktop computers backed by internet. They are called Computer Assisted Personal Interviewing (CAPI). The interviewees enter their data direct on the data base via their laptops or desktops. The most trusted and common strategy of gathering data via interview involve the direct conduct between the researcher and the interviewees. The strategy is popularly known as face to face interview.

From the philosophical perspective, face to face interview is ideal for qualitative gathering. According to the social constructionist paradigm (a philosophy behind qualitative research), there should be zero distance between the researcher and the researched (Creswell, 2014). The researchers are allowed to interact with the research subjects. The process of data gathering in qualitative research involves the interaction of the researcher and research subjects in the research site.

Qualitative interviews are unique, they are different from quantitative methods of data gathering by being open ended. They enable researchers to explore research problems in detail. Qualitative studies are said to be intensive and complex because the data is gathered by open ended instruments like interview guide as well as focus group guide. In-depth interviews are open ended.

There are several types of interview. Hancock (2002) talks of highly structured, semi structured and unstructured interviews. The duty of the qualitative researchers is to adopt types of interviews that satisfy the dictates of social constructionist philosophy and the nature of the study. The researchers should make sure that the interview guide has open ended questions.

Highly structured interviews give the interviewees limited space to express their views. They consist of closed ended questions. For example;

- *Interviewer: Do you think your president is performing poorly, well or excellent in the area of governance?*
- *Interviewee: The president is performing well.*

The interviewee is not given enough platform to argue out of the box because the interview session is predetermined. The interview has closed ended questions and the practice is against the dictates of qualitative philosophy. Highly structured interviews cannot fit to be under qualitative methods of data gathering basing on philosophical worldview of qualitative paradigm. The highly structured interview guide should be followed from the beginning of the interview to the end. The interviewees are not flexible to express their views.

The other type of interview is known as semi-structured. The semi-structured interviews contains both open and closed ended questions. They are sometimes called focused interviews. The interviewers are allowed to probe during the interview sections. The interviewees are also allowed to express their views freely. Both closed and open ended questions should be written down on paper before the interview session. For example;

- *Interviewer: Do you think your president is performing poorly, well or excellent in the area of governance?*
- *Interviewee: The president is performing well.*
- *Interviewer: Why do you think so?*
- *Interviewee: From my own encounters and experiences with this regime, there has been a great improvement on governance. The devolution is working properly and funds are given to the devolved units as per the dictates of the constitution ....*

The closed ended questions are used by interviewers to prepare the interviewees to answer the open ended questions. Semi structured interviews are enjoying a wide spread usage in qualitative studies. They are recommended for use in qualitative projects. The interviewees should be given enough space to share their opinions and experiences.

The third type of interview is called unstructured. It is often called in-depth or depth interview and is believed to be extensive and detailed. The unstructured interviews are said to be having little structure. The interviewer will only prepare topics that can guide him/her during the interview session. The researcher may use research objectives or questions as a guide during interview session. Most of the interview questions to be asked are largely open ended. The interviewees are asked questions based on topics. Just like the semi structured interviews, unstructured interviews are also enjoying wide spread usage in qualitative projects. They are proper instruments for qualitative projects.

The interview guide (instrument) should be divided into key sections. The interview guide is supposed to have an introduction, bio data, content and closing remarks. The researcher should introduce him/herself and issues related to ethical considerations should be considered as well. The bio data section should contain the demographic information of the research subjects. The content of the interview should be closely connected to the research objectives or research questions. The interviewer should conclude by thanking the interviewee for participating in the study. The example below shows a structure of an interview guide;

Critical Thinking Challenge

1. Highly structured interviews should not be considered under qualitative data gathering tool. What is your view?

Interview Guide for Graduate Students

Introduction: My name is Silas a PhD student of …… I am undertaking a study on the attitude of university students towards inferential statistics. Feel free to participate in this interview and you are assured of confidentiality. The interview will not take more than 45 minutes.

Gender .... Program .... Year of Study...............

Other Family Details (if relevant) ....

1. Have you studied statistics before?
2. What was your attitude towards inferential statistics before the training?
3. Did your attitude change after the training?
4. What do you think were the factors that influenced your attitude change after the training?
5. What do you think are the reasons as to why graduate students are not using inferential statistics in their research projects and their lives?
6. Were you interested in inferential statistics before the training? Why
7. Did your interest change after the training? Why
8. Did your attitude towards inferential statistics affect your performance in inferential statistics exam? Why
9. What are the challenges that you face when learning inferential statistics?
10. What do you think can be the possible solutions to the problems?
11. In what ways do instructors influence students in learning inferential statistics?

Thank you so much for your valuable contribution

The administration of interviews involve some steps that should be followed by the interviewer. The interviewer should begin by introducing him/herself and the ethical issue related to the study should be communicated to the interviewee. The interviewee is also given some tips on how the interview is supposed to be undertaken. The interviewers should also request for permission to use video or audio recording instruments if they intend to use them during data gathering. The question and answer session should begin. After the question and answer session, the interviewer should ask the interviewee to ask any questions before closing the interview sessions.

Critical Thinking Challenge

1. Select a research problem of your choice, design a semi structured interview guide?

Focus Group Discussion Method in Qualitative Research

Focus group discussion and interview method are closely related. The only difference is that interview method involves a single interviewee during interview session. On the other hand, focus group discussion session involves more than two interviewees. Focus group discussion sessions are extensive group interviews aiming at gathering information from a group of people with common characteristics. Just like the interviews, the focus group discussion guides should consist of open ended questions. The focus group discussion instruments should contain the following sections; introduction, bio data, content and closing remarks.

The focus group discussion method is used when resources are limited. Undertaking data gatherings with a group of people per session do not only save cost but also time. The other factor that my provoke researchers to use focus group discussion is related to the nature of the interviewees. The interviewees should have the same characteristics. Having similar characteristics mean that they should all participate in the focus group discussion session with minimum challenges.

The first step in gathering data via focus group discussion guide is for researchers to study the population from which research elements are selected. The interviewers should make sure that the entire population is represented during the focus group session. A session for focus group discussion should consist of a considerable number of interviewees who are manageable. The American Statistical Association (1997) suggested between 6-12 interviewees. Hancock (2002) suggests 6-10 interviewees per session. The number of interviewees should be manageable. A figure in assess of 15 interviewee may be difficult to manage.

There are some key issues that should be considered just before undertaking an interview with a group of people. The moderator of the group interview session should be highly experienced. If not experienced, the moderator should acquire some important skills before undertaking the interview sessions. The moderator and the research team should make use of proper methods of audio and video recording instruments. The recording instruments should be able to properly capture appropriate data from the interviewees. They should be backed by other instruments in case of malfunctions.

The moderator should be able to keep the conversation flowing. The interviewees should be motivated throughout the group interview sessions. Time keeping is also another component to consider. If the interview session is supposed to be finished within one hour, the moderator should make sure that the discussions are completed within a stipulated time. The moderator should remain neutral and each and every interviewee should be given equal time and space to contribute to the discussion. The outspoken interviewees should be neutralised so that every interviewee will get equal share of time. The moderator should also avoid unnecessary interferences that may compromise the quality of the discussion and the flow of idea. The duty of the moderator should be to moderate and not to be a contributor to the discussion.

The researchers should appreciate that administering focus group discussion interviews is a complex exercise. They have to be very careful on recording the group interview data. Every interviewee's contribution should be properly captured by the recording instruments. A disaster may arise if the recording instruments are not backed. The moderator should make sure that all interviews are given equal space to participate in the discussion. A credible focus group discussion is the one in which all elements participating in data gathering are given equal space to contribute on each question.

Observation Method in Qualitative Research

Observation is the systematic analysis of the research elements in their natural settings by the observer (researcher). The researchers/observers have to use their skills, intelligence, experience and senses to arrive at acceptable conclusions about the elements under the study. The observation method in qualitative research involves the interaction of the observer with the elements on the research sites and the recording of what has been observed (Gold, 1958).

According to the social constructionist philosophy that gave rise to qualitative paradigm in research, the observers and the observed have to interact during the stages of data gathering. Knowledge is generated during such interactions between the observer and the observed. In contrast, quantitative researchers call such interactions biases. They are against the practice of the interaction of the researcher/observer and the researched/observed. The interaction between the observer and the observed is sometimes necessary. It helps the observer to collect reliable and credible data from the research site.

There are several circumstances that should be considered in order for qualitative researchers to make use of the observation method. In most cases, observation methods may be used when answering questions like what and how. There may be need for observers to visit the research sites in order to observe how the phenomena are going own. According to Hancock (2002), the duty of qualitative research studies is to give answers to what and how questions. The researchers should visit the research site in order to observe how they (research elements) behave. Therefore, separating the observer from the observed may not be possible.

The use of observation as a data collection method may be considered if little information is known about the elements in the population. The researchers/observers should visit the research sites in order to interact and understand the elements of the population much better. The visits will enable the observers to know what is going on research sites. It may be difficult for the observer to have enough knowledge about the elements of the population if visits to the sites are not done.

The observation method of data gathering may be used for verification purposes. The observers may want to verify what is said and what is on the ground. It may be sometimes difficult to accept what is being said before verifying with what is happening on the ground. There is therefore need for qualitative researchers to visit sites and observe what is happening on the ground. The observers need to verify facts, undertaking qualitative research projects in absentia is not acceptable. The researchers should rely on primary data rather than secondary data alone.

There are two basic type of observation that qualitative researchers should consider (Savage, 2000). The qualitative researchers may opt for participant observation, non-participant observation or both. The selection of an observation method to adopt depends on the nature of the study and the natural settings of the research sites. Participant observation has a lot of advantages as compared to non-participant observation. However, the use of both types of observation methods is highly encouraged. This section did not consider covert observation because of its conflict with the qualitative ethics. Under covert observation, the observers goes underground. The population will not be aware on whether they are being observed. Qualitative research ethics are against such practice. The observer should identify her/himself and permission to undertake the study is granted.

Critical Challenge Question

1. Is it proper to undertake observation under cover in qualitative research? Why

Participant Observation

Participant observation is sometimes known as direct observation. The observer analyse research elements/phenomena/activities from within a population. The observers interact and participate in the activities of the group. Participant observation is enjoying widespread usage in qualitative research because of its practicability and effectiveness. However, the observers need to adhere to certain ethical conditions in order for the process of data gathering to be smooth. Overt-participant observation can be also used in place of participant observation.

The observer can be a member of the same group under the study or a person who grew up with the elements of the population. The observer will find it easy to undertake the study because of easy acceptability by the group members. The observer knows the language, culture and the life styles of the group. In case of studying the lifestyle of Hadzabe tribe in Northern Tanzania, selecting observers/researchers who are members of the Hadzabe tribe is advantageous. The Hadzabe observers are familiar with the culture of the population and they are also accepted by the fellow community members. A problem comes if the observers are not accepted by the population of the research site.

On the other hand, the observers may be from outside the communities under the study. However, the observers should be familiarized with the culture and the life styles of the elements at the research sites before undertaking their studies. In the case of a study of the lifestyle of Hadzabe tribe in Northern Tanzania, observers may be persons who are not members of the Hadzabe community. The observers should start by studying the lifestyles and culture of the Hadzabe people. They should also understand their language or make use of interpreters. The challenge of such observers is not only on acceptability but on language of communication. That is why local observers are preferred ahead of outside observers. An arrangement to engage interpreters should be made in advance.

In a nutshell, participant observation has been enjoying a wide spread usage in qualitative studies because of its effectiveness in gathering data. Observers learn better about the elements and the population while they are studying from within. The observers are acceptable to the communities, the observer and the observed relationship is always conducive for data gathering. The selection of observers from within the population under the study has a lot of advantages than selecting observers from outside the population under the study. Outside observers face communication challenges and the use of interpreters may be costly to the research project.

Non-Participant Observation

Non-participant observation is another type of observation method. It is sometimes called passive observation or indirect observation or overt-non participant observation. The observer observes from a distance and not from within the population. (Jorgensen, 1989). The observers should introduce themselves to the research subjects, they have to declare their interests, purpose of the study and ethics to be considered. It is different from covert observation were observers go underground. Qualitative research ethics are against under-cover observation. Permission and consent should be agreed by both parties.

Just like the participant observation, the observers should begin by studying the local culture and the life styles of the population under the study. They have to study their language of communication or engage interpreters. The observers may be from within the population and there are a lot of advantages connected to this. The local observers know and practice the language, culture and lifestyles of populations under the study. The selection of outside observers (from outside the population under the study) may have a lot of challenges and can be costly to the project. The use of non-participant observation has a lot of challenges, the observers may find it hard to penetrate the hearts of the population elements. The population elements may see observers as strangers. The research team should work hard to reduce or eliminate such challenges.

There are several techniques of gathering data during the observation. The observers may use the pen-paper method. It means that the observers should write down what they observe in their diaries or pamphlets designed for data gatherings. The observers may also use audio or video recorders but they should get permission from the elements involved. The observers may also collect samples or artefacts from the research sites for further tests but they should get permission from the people in charge of sites. They should avoid vandalizing the research sites or taking sample without written or verbal permission from the people in charge of the research sites.

There are ethics to consider when dealing with elements of the population during the process of data gathering. If tribes or people with common culture are involved, the observers should get permission from the gatekeepers before undertaking their studies. The gatekeepers should introduce the observers to the sites. They should also assist the observers to access the key informants.

The observers should also be concerned with the reactivity effect. The population under the study may react unfriendly and the observers should work with the gatekeepers in order to solve reactivity challenges. The observers should study the culture of the population under the study in advance. The observers may consider reciprocity. They may give some gifts to the elements involved in the study or the gatekeepers. They may also support in community projects if the funds are available. However, the gifts should not be used to spoil the elements. The elements should not be manipulated by the gifts. They should always give correct data to the observers.

The observation method of data gathering should make use of observation guides when collecting their data from the research sites. The observation guides make the observers to be focused. The observation guides have to contain information about the following;

- Introduction – names of the researchers, purpose and guarantees of confidentiality/privacy.
- Bio data – the demographic characteristics of the elements participating in the study
- Content – what is being observed? Are they studying the language, culture, life style, artefacts, marriage practices, food preferences or health services.
- Observations and remarks – there should be a section containing what was observed in the field during the gathering of data. The observations should be made based on the objective of the study. The observers should give their own remarks about what has been observed.
- Findings – the observers should make findings on what they observed.
- Conclusions – the findings should lead the observers to the conclusion or conclusions depending on the objectives of the study

Critical Thinking Challenge

1. Select a problem/title of your choice and design an observation guide to be used for data gathering?

Documentary Analysis in Qualitative Research

Documents are key sources of data in qualitative research. Documentary analysis is a process of data gathering and interpretation which enables the researchers to make decisions in qualitative projects about documents under the study (Bowen, 2009). The documents cannot be avoided in qualitative research because they contain vital information about history. Documentary analysis is used across qualitative designs but the method is popular with historical designs.

Documentary analysis is important in qualitative research projects in several ways. Documents keep researchers and historians alive with the past. The researchers need to study what happened in the past because the future is often controlled by the past. The documents preserve information and protect it from extinction. Researchers can only be able to preserve information by keeping it safe in documents. Documents are also important because they contain details about traditions and how they evolve overtime. There are quite a number of researchers interested in studying how different culture evolved overtime. They have to make use of documents reserved. This section educates readers on the processes of analysing documents in qualitative projects.

There are two popular documents in qualitative research. There are primary and secondary documents. Both primary and secondary documents are valuable in qualitative studies. The duty of the researchers is to try and separate secondary from primary documents. Primary documents are superior to secondary documents because they contain live information from eye witnesses or persons who participated in the phenomena. On the other hand, secondary documents contain information from the persons who heard from the eye witnesses or those who participated in the events. In short, primary documents and secondary materials can be similar but the difference is based on whether they are a reflection of what the participants say or what the eye witnesses saw or whether they contain information from the persons who heard from the eye witnesses or participants.

Primary Data Documents in Qualitative Research

For a document to qualify to be primary, it should contain information direct from the participants or information from the persons who witnessed the events or the phenomena. The following can be part of primary documents;

constitutions, personal diaries, court decisions, letters, emails, wills, affidavits, certificates, bills, receipts, maps, verdicts, pictures, paintings and research reports. Such documents qualify to be primary if they contain information from the participants or eye witnesses (persons who saw it happening). They seizes to be primary if they are reports from those who heard from the participants or eye witnesses.

There is always confusion among researchers when handling primary and secondary data sources. Some think that interviews are part of primary data collection sources while document analysis/sources are secondary. Documents of history can be secondary or primary. They are primary documents if they contain information from the participants or eye witnesses. They can be secondary documents if they contain information from persons who heard from participants or eye witnesses. In qualitative research, interviews, focus group discussion and observation methods are not questionable primary data sources if they are undertaken directly or face to face with the research subjects.

According to O'Leary (2014), primary documents are classified into three categories. The first category consists of public official documents of organizations i.e. annual reports, registers, employee records, budgets, minutes of meetings and other documents of the organizations. Such documents should be generated by individual organizations and they are classified as primary documents.

The second category of primary documents consists of personal documents. Personal documents are documents that are linked to an individual and not organizations. Some examples of personal documents are; certificates, textbooks, journal articles, emails, letters, autobiographies and social media related documents like Facebook, WhatsApp, twitter accounts and others. Such kinds of documents are useful in qualitative research. The personal documents should be generated by an individual directly in order for them to be classified as primary documents. There should be enough evidence that the documents were compiled by an individual. Documents like wills are often challenged in courts of law because some persons may doubt their originality. They may not be genuine primary documents. The duty of the courts is to check whether they are genuine primary documents.

The last category of primary documents according to O'Leary (2014) is classified as physical evidence or documents. These are documents that exist in their natural state and not in their artificial state. There are some controversies that may arise when identifying such documents or evidence. For example, should we classify a mountain that we see on photography or video as part of primary document? Should we classify photographs or video as primary documents? Primary documents should be in their original state and examples of primary documents are artefacts, flayers, posters, agenda, transcripts, photographs, audio and video related documents.

Classification of primary documents can be confusing but the researchers should note that primary document may be linked to participants or eye witnesses. The primary documents may be physically available and can be observed by our sense i.e. touching, seeing, and hearing. A photographs, video and audio recorders can be primary because we see and hear them. Flayers and posters can be primary because see, read and touch them. The author strongly believe that the criteria for identify documents should be based on three senses i.e. seeing, touching and hearing. The other senses like smelling and feeling may not help us to identify documents. Qualitative researchers should be very careful when identifying primary and secondary research documents.

Secondary Data Documents in Qualitative Research

Secondary documents are important in qualitative research and are enjoying a wide spread usage. It is important for readers to note that secondary documents in qualitative research are not original. They are not original because they are not directly connected to the participants or the persons who directly observed the phenomena/events (eye witnesses). A good example of secondary data are journal articles that are classified under empirical literatures. Some researchers may accomplish their studies online supported by empirical literature without going to the research sites. Such kinds of literature documents used may be classified as secondary documents.

Emails, letters, court decisions and textbooks may be secondary data if they are not directly linked to the elements under the study but can be primary if they are products of the elements drawn from the population. In some cases, the persons from the research sites may produce documents like emails, letters, manuals and pamphlets on research site and such documents may be classified as primary.

The qualitative researchers should be able to make a distinction between primary and secondary documents. The criteria for distinction should be based on whether such documents are linked to participants or eye witnesses and the basis of distinction may be based on three senses i.e. touching, hearing and seeing. The qualitative researchers should also note that feeling and smelling may not be able to identify documents. It should be also noted that every document may become primary or secondary depending with contexts. The notion that emails and letters are always primary documents is a mistake because they may be secondary documents to other contexts.

Steps of Analysing Documents

There are several steps that may be followed when analysing documents but the steps may not be limited to the ones suggested below.

Identification of documents – the first step for researchers is to identify the documents to observe from the population or linked to the population. There may be need for the sampling of such documents if they are many. The documents should be selected from the entire population. They should be able to be a representative sample of the population. Furthermore, the documents selected should be relevant to the objectives of the study, only valid and relevant documents have to be considered.

Analysis of documents – after the identification of documents, the analysis of documents should begin but the analysis should be based on the objectives of the study. All the information from the documents should be jotted down and a pen-paper method may be used to write down what is contained in documents. The qualitative researchers should take time listening/observing audio/videos/photographs and writing down what they observe from these the documents.

Transcription of data – the data obtained from each document should be transcribed. The transcription may be verbatim or non-verbatim. The transcription should be based on the nature and demands of the study. This book contains detailed information about transcription and readers should take their time going through the details of transcription in chapter eight.

Coding of data – the data from the transcription should be categorised. The researchers are expected to perform open coding. However, axial and selective coding may be performed outside the grounded theory. The codes should be used to represent categories and this should be done according to the rules of analysing qualitative data.

Report writing – after the coding of data into categories or themes, the researchers may start to compile their reports. Their reports should be able to address the objectives of the study and the researchers should avoid biases. The information from the documents should be heard and not the voice of the researchers. The researchers should follow ethics for research reports writing. The rule of confidentiality should be observed. Names of individuals or organizations may be concealed if there is need.

The readers should by now appreciate four major data gathering methods in qualitative research. Data gathering methods like interview, focus group discussion, observation and document analysis are popular in qualitative research projects. The researchers should follow the basic steps of gathering data by these methods. Research ethics should be also considered in order to protect research subjects from harm. The confidentiality and privacy rules should be strictly observed by qualitative researchers.

Critical Thinking Challenge

1. Describe the differences between primary secondary documents?
2. State and justify ethics connected to data gathering stage?

3 QUESTION AND ANSWER SESSION

What is the Difference Between Qualitative and Quantitative Data Collection Methods and Instruments?

The key difference between qualitative and quantitative data collection methods and instruments is that, the qualitative instruments and methods are open ended while quantitative data collection instruments and methods are closed ended. In short, quantitative instruments are predetermined, respondents have no freedom of thinking outside the box. They are given questions and option of answers to select from.

The following are some of the data collection methods for quantitative projects; closed ended questionnaires, closed ended scales, closed ended tests and closed ended checklists. There are four major types of qualitative data collection methods that are enjoying widespread usage in qualitative research projects. These are; interview, documents, observation and focus group discussion.

What are the Types of Interviews?

There are several types of interviews but there are three key types of interviews. The first type of interview is called highly structured interview.

The type of questions are closed ended and are questioned for not allowing respondents to express their views outside the response options given. This kind of interview may not be considered to be under qualitative data collection methods. The second type of interview is called semi structured interview, it consists of some closed ended questions and a lot of open ended questions. All the interview questions are written down in advance and administered to the interviewees. The interviewers will follow the interview guide till the end of the interview session.

The third type of an interview is called unstructured interview. They are often called depth or in-depth interviews. There is no clear structure under this type of interview. The interviewer will prepare project objective/questions/ topics and call for an interview with the interviewee. The interviewer will use the topics/objectives to ask questions. However, semi structured and structured interviews are considered to be the typical types of qualitative interview because of their use of open ended questions.

What is the Difference Between Interview and Focus Group Discussion Methods?

It is true that focus group discussion interviews are different from the interview method. The focus group discussion is sometimes called group interviews because they are done with two or more interviewees. Interview are done with a single interviewee.

When Should Researchers Prefer Focus Group Discussion Ahead of Interview Method?

The focus group discussion can be used to save cost and time. Interviewing a group of persons at the same time is not only cost effective but also good for managing time. The focus group discussion method may be used when there is a shortage of resources and time. The focus group discussion can be also used when the people to be interviewed have similar characteristics. They have to share common characteristics.

How many Persons Should be Involved in a Single Group Interview Session (Focus Group Discussion)?

Group interview means that there should be two or more people to participate in interviews. However, according to American Statistical Association (1997) the number of persons for group interview should be 6-12 persons. On the other hand, Hancock (2002) suggested 6-10 persons to be part of the interview. The number of persons per interview session should be manageable. Huge numbers of persons per session may disturb the flow of the group interview processes.

What are the key Issues to Consider During Group Interview?

The interviewer have to be neutral and each and every participant should be given balanced opportunity to participate in the group interview. The interviewer should also consider timing. The interview should be finished within a stipulated time. The interviewers are supposed to acquire relevant skills of handling group interviews.

What is Observation Method?

In some cases, the researchers (observers) have to visit the research sites in order to verify facts and experience what is happening on the ground. Observation method is a technique of assessing the research elements in their natural setting. The observers have to visit research sites and have a face to face interaction with elements on research sites. The method is very important because it can be used for verification purposes. The method can be used when little information about elements is available and when questions like how and what are supposed to be answered.

What are the Types of Observation?

There are several types of observation method, the types of observation can be participant and non-participant observation. Under participant observation, the observer observes from within the group under the study. The observer can be a member of the group, grew up with them, practice the same culture and speak the same language. However, an observer outside the group may also come in and study from within but should have to learn their culture, language and their life style. Interpreters may be used if there is language challenge. All in all participant observation is appropriate because the observers are accepted with the members of the groups and there is little friction. The challenges are mainly of bracketing the observers' experiences, emotions and feelings.

Non-participant observation is the opposite of participant observation. The observers observes from a distance and often called passive observation. The major challenge of this method is on acceptability by the group under the study. The members of the group may not trust the observer leading to the gathering of invalid information. The method is less effective compared to participant observation. It is always effective to fight from within rather than from outside.

What is Document Analysis?

Documents are written, audio and video records that contain information of the past. Qualitative researchers often making use of documents. Documents contain vital information of the past. They preserve traditions and keep the historians linked with the past. The documents are divided into primary and secondary documents. However, both primary and secondary documents are important in qualitative research though primary documents have a lot of advantages.

What is the Difference Between Primary and Secondary Documents?

It is sometimes difficult to differentiate between primary and secondary documents. The first criterion of distinction may be based on whether the document was authored by the participant. If the document is authored by the participant, it becomes a primary document. If it the document is authored by an eye witness or a person who saw it happening, it becomes a primary document as well. On the other hand, if the document is not related to the participants or the eye witnesses it becomes a secondary document. Primary documents are considered to be original and genuine.

What are Primary or Secondary Data Collection Methods in Research?

There is always confusion among the researchers, some think that, interview and focus group discussion alone are primary documents while documentary analysis is classified under secondary data collection methods. Interviews, focus group discussion and documentary analysis can be primary or secondary documents depending with the criteria set above (under section 4.2.10). The documentary analysis method can be primary or secondary data collection method depending with the criteria set above.

REFERENCES

Bogdan, R., & Biklen, S. (2007). Qualitative Research for Education: A introduction. In Research Design. Qualitative, Quantitative and Mixed Methods Approach (2nd ed.). Thousand Oaks, CA: Sage Publication.

Creswell, J., & Morgan, D. (1996). *Qualitative Research Method Series* (2nd ed.). Sage Production.

Denzin, N. K., & Lincoln, Y. S. (Eds.). (2005). *The Sage Handbook of Qualitative Research* (3rd ed.). Sage.

Glaser, B. G., & Strauss, A. (1967). *Discovery of grounded theory*. Aldine.

Gold, R. L. (1958). Roles in sociological field observations. *Social Forces*, *36*(3), 217–223. doi:10.2307/2573808

Guba, E. G., & Lincoln, Y. S. (2005). Paradigmatic Controversies Contradictions and Emerging Confluences. In N. K. Denzin & Y. S. Lincoln (Eds.), *The Sage Handbook of Qualitative Research* (3rd ed., pp. 191–215). Sage.

Jorgensen, D. L. (1989). Applied social research methods series: Vol. 15. *Participant Observation: A methodology for human studies*. Sage Publications, Inc. doi:10.4135/9781412985376

Locke, L. F., Silverman, S. J., & Spirduso, W. W. (2010). *Reading and Understanding Research* (3rd ed.). Sage Publication.

Marshall, C., & Rossman, G. B. (2010). *Designing Qualitative Research* (5th ed.). Sage Publication.

McCracken, G. (1988). *The Long Interview. Qualitative Research Methods series 13*. Sage.

Merriam, S. B. (1998). *Qualitative Research and Case Study Applications in Education*. Academic Press.

Morse, J. M. (2006). The Politics of Developing Research Methods. *Qualitative Health Research*, *16*(1), 3–4.

O'Leary, Z. (2014). *The Essential Guide to doing your Research Project* (2nd ed.).

Pitney, W. A., & Parker, J. (2009). *Qualitative Research in Physical Activity and the Health Professions*. Academic Press.

Savage, J. (2000). Observation: Standing in the Shoes of Others. *Qualitative Health Research*, *10*(3), 324–339. doi:10.1177/104973200129118471 PMID:10947479

Chapter 5
Qualitative Sampling Methods

ABSTRACT

The chapter takes the readers through five key types of non-probability procedures and is divided into six sections. The first section addresses purposive sampling and its seven variants. The variants discussed are maximum variation, homogeneous, total population, expert, critical case study, deviant case, and typical case sampling. The second section contains information related to quota sampling procedure and its differences with stratified and cluster sampling. The convenience sampling procedure is discussed in Section 3, snowball sampling in Section 4, and self-selective sampling in Section 5. The chapter is concluded by a question and answer section, which contains questions summarizing the entire chapter.

1 KEY CONCEPTS

- **Sampling –** selection of respondents/elements from a population.
- **Population** – total number of persons in a research site (where a project is to be undertaken).
- **Sample** – elements/respondents selected from a population that are supposed to directly participate in a project.
- **Subjectivity** – the excessive interference of a researcher in qualitative projects/bias in qualitative projects.
- **Objectivity** – control of bias in qualitative projects i.e. bracketing of the researcher's feelings in qualitative reports.

DOI: 10.4018/978-1-7998-8549-8.ch005

- **Probability Sampling** – sampling procedures controlled by randomisation where all elements have equal chance of selection to be part of the study.
- **Non-Probability Sampling** – sampling procedures in which the selection of respondents is controlled by the researchers (elements have no equal chances of selection).

2 LEARNING OUTCOMES

By the end of the chapter, the readers should be able to;

- Identify variants for purposive sampling procedure
- Understand steps of qualitative sampling
- Describe popular qualitative sampling methods
- Apply qualitative sampling methods in projects

3 CASE STUDY

Mr. Silas was a team leader of a qualitative project dealing with HIV patients in Buhera District. The team initially decided to adopt random sampling procedure when picking samples from the population. Unfortunately, the project did not proceed well because the team failed to identify enough samples for the project. The team had to engage experts for advice. They were advised to adopt a sampling procedure from non-probability family. The team decided to switch to snowball sampling and they managed to get enough samples for the project. By the end of this chapter, you should be able to appreciate that non-probability sampling procedures can be used to locate hidden population or samples.

4 INTRODUCTION

The case study above should teach readers some important lessons on sampling. The use of wrong sampling procedures may be costly to a project. The project team should pick appropriate sampling procedures in order to avoid unnecessary inconveniences. The positivists and social constructionists hold different views on sampling procedures. Research studies/projects of qualitative and quantitative are done with representative samples selected from a population. A doctor testing malaria will start by collecting a sample of some drops of blood from a patient. There is no need for a doctor to take litres of blood from a patient. The same applies to social research studies, researchers gather their data from few samples selected from a population. The main aim of this section is to enable researchers and readers to apply non-probability sampling procedures in qualitative research projects. This section is divided into the following headings;

- Purposive Sampling Procedures
- Quota Sampling Procedure
- Convenience Sampling Procedure
- Snowball Sampling Procedure
- Self-Selective Sampling Procedure
- Question and Answer Session

Qualitative specialists make use of non-probability sampling procedures when undertaking their projects. On the other hand quantitative specialists make use of probability sampling procedures when undertaking their studies/ projects (Wretman, 2010). Non-probability sampling procedures are questioned for being subjective (Chaudhuri & Stenger, 2005). The process of selecting samples from the population is controlled by the researcher. The quantitative specialists believe that such kind of researcher's interference with the selection of research subjects is a true example of bias. They believe that the sampling procedures for qualitative are not objective but biased.

The interferences of researchers with the sampling process is normal in the perspective of the qualitative researchers. According to the social constructionists' worldview, it is not possible to separate the researchers from the research subjects (Creswell & Clarke, 2007). Both researchers and research subjects have a role to play during the process of undertaking qualitative research projects. The quantitative researchers view non-probability sampling procedures as weak methods of sampling elements from a population.

The other tenet of non-probability sampling procedures is that they are used to select samples from a small population. They are only applicable when handling research projects with few research subjects. The probability sampling procedures are used to select samples from a large population. The implication to practice is that non-probability sampling procedures are used to select few samples while the probability samples are used to select large samples from the population.

There are several reasons for using non-probability sampling procedures when undertaking research projects. Shorten & Mooley (2014) argued that non-probability sampling procedures may be used when it is impossible to use probability sampling procedures. It is impossible to use probability sampling when dealing with very few elements within a population. There are some qualitative designs like biography that deal with only one subject and making use of probability sampling will not be possible. There are also some research projects that are done on purposes that are critical that may not need the use of probability sampling procedures.

5 CONTROVERSIAL SAMPLING PROCEDURES

There are some sampling procedures that are controversial. It is sometimes difficult to classify them. They can be used under probability or non-probability sampling procedure. It is very difficult to classify systematic sampling. Bluman (2007) classify systematic under probability sampling. However, the challenge is that only one (the first) sample is selected via randomization while the rest of the samples are predetermined. It is therefore difficult to classify systematic sampling under probability (Madondo, 2016).

The other confusion is on stratified sampling and quota sampling. The quotas and strata are almost the same. There is no clear distinction between the two and they are often taken as neutral samples because they fit on both sides. According to Bluman (2007), under stratified sampling, randomization is done when selecting samples from the segments but segments are not identified by randomization. In cluster sampling, the selection of a cluster/strata is done via randomisation and all samples within a cluster will automatically become part of the study. There is no randomisation performed under quota sampling. The entire process is controlled by the researchers.

Both cluster, quota and stratified are characterised by segments but they are considered to be from different paradigms. However, probability and non-probability sampling procedures are all important in research. There are circumstances when probability sampling procedures are difficult to implement and non-probability samples may be used. The argument from the positivists that non-probability samples are weak may be a radical view. The pragmatic approach to sampling is of essence. The use of both probability and non-probability sampling procedures is likely to produce detailed and credible results.

Critical Thinking Challenge

1. Make an assessment of stratified, cluster and quota sampling procedures?

6 PURPOSIVE SAMPLING PROCEDURES

There are several terms that are used in place of purposive sampling. The following terms may be used in place of purposive sampling; purposeful, judgemental, selective or subjective sampling. The term subjective is used because the researchers play a key role in the selection of samples from the population (Zhi, 2014). The selection of the research subjects is based on purpose. The major challenge of the purposive sampling just like other non-probability sampling procedures is that the selected sample may not be the ideal representation of the entire population.

There are several variants/types of purposive sampling procedures. This section will address heterogeneous, homogeneous, typical case, extreme case, critical case, expert and comprehensive sampling procedures. The researchers should be extremely cautious when selecting samples. They need to check on whether the type of purposive sampling procedures adopted represents the entire population. The characters of the units of a population have to be considered when selecting the type of purposive sampling to adopt. The researchers need to understand the purpose of the study as well.

6.1 Maximum Variation Sampling

This type of purposive sampling is sometimes known as heterogeneous sampling. The aim of the sampling procedure is to capture different traits and characteristics of a population. The samples in a population are sometimes different in terms of their characteristics and behaviour. The duty of the researchers is to identify whether different characteristics exist within units or a population. After the identification of various characteristics, the researchers should go ahead selecting samples from a population based on characteristics identified.

In a nutshell, the researchers making use of maximum variation sampling should follow two basic steps. The first step is to identify different characteristics of elements within a population. These characteristics should control the process of sample selection. After the identification of different characteristics in a population, the researchers may go ahead picking their research subjects basing on the characteristics observed. The selection of samples may be proportionally or not proportionally represented. What should be considered is that all the different characteristics of the population are represented.

A researcher may wish to undertake a qualitative project on the attitudes of university students towards inferential statistics. The project is to be undertaken with a university consisting of 1200 students. The first step is for the team of researchers to identify the variations from the university population of 1200. The variations may be based on study disciplines. For example the researchers may want to categorize the population based on faculties (education, social science, natural sciences, theology, business etc.). The classification may also be based on study programs. This is done in order to ensure equal and fair representation of the entire population.

After the identification of variations, the researchers may start the selection of the samples based on the faculties selected. If a proportional representation is to be used, the selection of samples should be determined by the number of students per faculty. It means that, those faculties with high population will also get the highest number of samples. In some cases, disproportional representation may be used, equal number of samples will be selected from each faculty. The methods is used when comparison is to be made between units of characteristics.

The selection of sample size may be controlled by the population, availability of resources, time and duration of the study. The researchers have to play with such factors in order to make a decision on the sample size. The qualitative researchers are not concerned with the quantity but rather the feelings, emotions and experiences of research subjects representing units or various characteristics of the population.

6.2 Homogeneous Sampling

Maximum variation sampling or heterogeneous sampling is the opposite of the homogeneous sampling. The variation of characteristics within a population is not considered when dealing with homogeneous sampling procedure. The procedure is used when the characteristics of units within a population are similar. There will not be need for researchers to think of identifying differences within a population.

The researcher may want to investigate the food preferences of tuberculosis patients. The study is to be undertaken with a population of 34 tuberculosis patients at hospital B. The first step is for the researchers to come up with a sample size and select the research subjects from the 34 tuberculosis patients. The homogeneous sampling is proper to use because the population (of the 34 patients) consists of all tuberculosis patients. The similarity in terms of the characteristics of the population elements makes it possible for the researchers to adopt homogeneous sampling.

The homogeneous sampling is not complicated when selecting respondents or research subjects because all respondents within the population are capable of representing the entire population. The heterogeneous sampling can only be used if the population units are similar in terms of their characteristics and traits. If the population under the study has different characteristics, maximum variation may be considered. The researchers should therefore begin by critically analysing the units within a population. The purpose of analysis is to check for variations or similarities in terms of the characteristics of units within a population.

6.3 Total Population Sampling

Another type of purposive sampling is called total population sampling. This type of sampling is sometimes called comprehensive sampling. It is a qualitative sampling procedure in which all the elements of the population automatically become part of the sample size. Qualitative projects are not concerned with the bigger sample sizes. Their concern is what the individual subjects say and not the majority's views.

The key characteristic of total population sampling is that the population (from which the subjects are selected) should consist of very few elements for example, the entire population may be having less than 30 people or elements. *A good example is the study of the needs of blind people in village X. In most cases, the total population of blind people in the villages may be less than five people. The researcher is encouraged to study all units within a population.*

A researcher may also want to examine the food preferences of breast cancer patients in hospital C. If the research team is able to undertake the project with all the breast cancer victims at hospital C then comprehensive sampling procedure should be used to select samples. There will be no need for researchers to sample subjects from smaller populations. The advantage of comprehensive sampling is that, it is an ideal representation of the population because every element is captured to be part of the study. The sampling procedure is also cost effective because it deals with very small populations.

The total population sampling is the purest sampling of all probability and non-probability sampling procedures. Its use depends on the ability of the researchers to study all the units of the population. The population size may be high but comprehensive sampling procedure may still be used if the research team has enough resources and time to execute the task. Another example were comprehensive sampling is used is that of National Census. All people in a country should be counted. Therefore, the total population sampling is a function of time and available resources. The research project should be able to capture all the elements available in a population to be part of the study.

6.4 Expert Sampling

This type of purposive sampling is unique because it deals with professional people only. Professional people are experts in different disciplines. These are studies that should be undertaken with professional people as research subjects. The sampling procedure is used to pick research subjects who are knowledgeable in the area under the study.

A researcher may want to study the implementation of Paris Global Climatic Change Deal. Such an area requires experts in the field of climate change and the researchers should select climate change experts to be participants of the research project. The key task of the research team is to select key informants appropriate for the study. Researchers are requested to crosscheck on whether respondents are key informants. They can do that by asking the respondents to confirm before interviews or data gatherings proceed.

In some cases, the researchers should think of involving the services of experts to be part of the research team. It may be difficult for researchers who are not experts in the area under investigation to be part of the panellists or data gathering team. Therefore, the research team should be led by experts who understand the professional jargon and the language of the discipline.

Expert sampling is very useful when there is lack of empirical evidence in the area under the study. There are some areas of studies that are not backed with empirical evidence like the studies related to Ebola disease. The researchers under such areas may also capitalise on expert sampling. Expert sampling is easy to implement during data gathering because it involves experts. The problem may arise when experts are dealing with research team members that are not knowledgeable in the area. The purpose of expert sampling procedure is to select subjects who are experts in the area under the study and the research team should be very careful when selecting the key informants.

6.5 Critical Case Study

This sampling procedure is critical in the sense that it involves the selection of few and quality key informants from a bigger population. The sampling method is mainly used when there is a challenge of resources and shortage of time to undertake a study with a bigger sample. Because of the shortages to resources, time or the urgency to complete the study, the research team is forced to select quality and key informants who are classified as 'critical informants'. They are critical because they are able to provide responses that can represent the entire population.

The challenge to the team of researchers will be only on the selection of such quality and key informants. *For example, United Nations may want to undertake a pilot study ahead of the emergence food distribution project in country Y. The research team may consider making use of critical case sampling to select quality and key informants. The use of critical case sampling will facilitate the emergence distribution of food by saving time. The team of researchers may involve the NGOs working in the area in question and the top government officials in the area. They will be able to complete the study in a short period of time.*

Under normal circumstances, the team of the researchers could have selected samples from the area affected. However, such a move would have delayed the food distribution to vulnerable people. The used of critical case will be good in order to manage time for the emergence food distribution properly. In any kind of studies, the researchers may opt for a critical case sampling in order to manage time and resources effectively. This is only done by selecting few 'quality and key informants' who are able to provide conclusive and reliable information about the problem under the study.

The major challenge of critical case sampling is on the selection of the quality key informants who are able to substitute a much bigger sample size that could have been selected. The research team should target few informants who are able to provide the right information. Such informants should be able to provide empirical and genuine information about the problem under the study. If the study manages to select the right informants, proper and reliable conclusions will be made within a short space of time. The method is not only time effective but also cost effective.

6.6 Deviant Case Sampling

This type of purposive sampling is often called extreme case sampling. The deviant case sampling focuses on the selection of samples that are unusual or special. Such kinds of samples are sometimes called outliers because they deviate from the average. The samples may be also classified as abnormal or gifted/talented. *A good example is that of the selection of HIV patients that are becoming resistant to anti-retroviral drugs. The research team may want to explore why they are resistant to anti-retroviral drugs. These subjects are classified as deviant because they are not responding to the drug as per the expectation.*

A team of researchers may want to examine the performance of students in mathematics in District X. The focus may be given to those schools with zero percent pass rate in mathematics or those schools with 100% pass rates. Zero and 100% pass rates are unusual and special cases respectively. The selection of sample from a population should consider such unusual and special cases. The purpose of a deviant case sampling procedure is to target unusual or special samples.

There are some few steps that the research team should follow when dealing with deviant case sampling. The first step is to study the population. The aim of studying population is to identify samples or units that are unusual or special. If the samples in a population are usual or normal, then deviant case sampling will not be relevant and appropriate. After the identification of deviant cases within a population, the researchers may then go ahead selecting unusual or special samples to be part of the study.

The use of deviant case sampling procedure will enable the researcher to examine unusual or special cases. Research is all about solving the problems existing in the universe and the selection of deviant case sampling will help researchers to focus on unusual cases that may need attention. Deviant case sampling is a problem solving sampling approach that seeks to examine unusual or extreme cases that my need attention.

6.7 Typical Case Sampling

Typical case sampling is the opposite of deviant case sampling. While the focus of deviant case sampling is on unusual or special cases, typical case sampling focuses on average or normal cases. There are some few steps that the researchers might need to consider when handling typical case sampling. The first step is for the research team to study the population that should be under the study. The aim of examining the population is to identify average or normal cases within the population. The researchers can then select samples from the average population to be part of the study.

The research team my want to examine the livelihood of families surviving on less than dollar income per day. The duty of the research team will be to select such respondents from the population to be part of the study. Typical sampling may be used if the majority of people in a population are surviving on less than one dollar per day. The research team may also want to examine challenges facing families serving on average income. Such kinds of families should be selected from the population.

Purposive sampling procedures are unique because they are undertaken on purpose. The researchers should begin by identifying the purpose of the qualitative study before proceeding to select respondents to be part of the study. Purposive sampling procedures are cost and time effective but questions of subjectivity are still becoming a challenge to the credibility of the qualitative research findings. Researchers should therefore be careful with their level of interference when selecting research respondents. Their interferences should not affect the credibility of the study.

Critical Thinking Challenge

1. Identify conditions that may force qualitative researchers to adopt different types of purposive sampling methods? Give examples

7 QUOTA SAMPLING PROCEDURE

Quota sampling is among the types of non-probability sampling tests. Just like stratified and cluster sampling, quota sampling is controversial. It is closely related to stratified and cluster sampling procedures. According to Bluman (2007), stratified and cluster vary in their levels of random intervention. When dealing with stratified, the segmentation of population into units/segment is done first and randomization is then used to select units in each segment. Under cluster sampling, randomization is used to select a single segment and the segment is called a cluster. All elements of a cluster or the selected segment will automatically become part of the sample size.

Quota sampling is similar to stratified and cluster in that they are all involved in segmentation. Segments are called strata, cluster and quota in stratified, cluster and quota sampling procedures respectively. The only difference is that randomization is not in any processes of quota sampling procedure. According to Daniel (2012), quota sampling should consist of elements represented by specific characteristics. In politics, there are quotas for women representation in government or parliament. The same applies to some research projects. The researchers may come up with quotas depending with the nature of the studies.

There are some key steps that the research team should follow when selecting samples via quota sampling procedure. The first step is to study the population and identify the strata. The second step would be to compute the quota of each stratum. The computation may be based on equal representation of elements in all strata. If the sample size is 60 from 10 strata/segments, it means that each stratum should have six elements. The selection of quotas may be based on proportional representation. The implication is that the number of elements selected from each stratum/segment depends on the total units within each segment/stratum. The selection via proportional representation is called proportional quota sampling. The last step would be to select samples from every stratum until the sample size is exhausted.

In case of undertaking a project on the attitudes of university students towards mathematics. The research team may begin by analysing the population of the university. The analysis should give rise to quotas and the quotas may be based on faculties. Each and every faculty should be given a quota i.e. business, natural sciences, social sciences, humanities, education etc. The selection of quotas may be based on equal representation or proportional representation. Once the quotas are computed, the research team may proceed to pick informants from each segment.

The quota sampling is hailed for pushing for the representation of all elements in a population. A credible study is the one in which all elements of the population are represented. Quota sampling has shown positive results in politics and research projects undertaken across disciplines. Quota sampling brings joy and satisfaction within a population and researchers should make use of this sampling procedure if need arises.

On the other hand, a lot of questions are being raised on the credibility of quota sampling. The strategy has been challenged for being subjective (Chaudhuri & Stenger, 2005). There is no randomisation under quota sampling. The quantitative research specialists believe that the process of sample selection is biased. There is a lot of interference by the researchers and the credibility of quota sampling will not satisfy quantitative researchers as long as randomization is not done during the processes of subjects' selection from the population. Despite the questions raised on the objectivity of quota sampling, the method is enjoying widespread usage in evaluation research of bigger projects. However, a pragmatic approach of handling quota sampling may be necessary it works and produce desirable results. There won't be problem if randomisation can be introduced when dealing with situations that require objectivity.

8 SNOWBALL SAMPLING PROCEDURE

According to Doherty (1994), there are situations that may provoke the use of snowball sampling procedure in qualitative research projects. The method of selection may be used when the population is hidden. There are some studies that deal with hidden populations and the research team should trace them to the hibernation sites via snowball sampling procedure. Snowball is the proper sampling procedure to use when the population is not accessible.

There are some key steps that should be followed when dealing with snowball sampling. The first step is to identify the initial sample or few samples. Accessing the initial sample or few samples may be tiresome and hectic. Once the initial sample or few samples are identified, those samples (few research subjects) should led the researchers to more samples that are hidden. Every element that participates in the study should recommend new samples. The strategy will continue till the research team gets enough samples for their project.

The other area to consider on snowball is the issue of research ethics. In most cases, the research subject are in hibernation because of privacy and confidential concerns. The research team should give assurance of confidentiality and privacy to the research subjects. In most cases, the process of data gathering should be done underground. The respondents should not be exposed to the public because they may want to remain in private domain. The research team may have face to face discussions with the research subjects in closed environments. This may help them to open up during the period of data gathering. The reciprocity tradition may be extended to the subjects involved in the study. The research team may assist the research subjects with medicine and other medical facilities if the resources are available. However, the research team should avoid spoiling the research subjects. Spoiling them may result in the provision of fake information.

A research team may want to explore the challenges facing people living with HIV in area Z. Most of the HIV victims are in hibernation, they are not comfortable to expose their status because of fear of being segregated. The research team may look for a sample of very few HIV patients. The identified patients will then assist them to identify additional subjects to be part of the study. The research team should give assurance on the respect for privacy and confidentiality. The process of data gathering should be done under closed environments and the respondents should be protected from the public eye. In some cases, if a research team has enough resources, they may help the victims with medical facilities. This will improve the researcher – participant relationship.

It is very difficult for probability sampling procedures to pick samples from hidden populations. Quota sampling can be among the suitable methods that can be used to access a population that is in hibernation. It is therefore mandatory for researchers to note that confidentiality and privacy rules should be handled with care when dealing with a population that is in hibernation. The processes of data gathering should be done in privacy especially when researchers are dealing with issues that are sensitive. The studies dealing with drug addicts may also use snowball to select research subjects, which is why confidentiality and privacy rules should be handled with maximum care.

Critical Thinking Challenge

1. Describe and explain ethical issues connected to snowball sampling?

9 CONVENIENCE SAMPLING

Another popular non-probability sampling procedure is convenience sampling. It is a sampling procedure in which samples are selected based on availability (Gravetters & Fornazo, 2012). The strategy is easy to implement and the research team should observe some simple steps that should lead them to the research subjects.

The first step is for the research team to follow is to observe and identify the population. The population should be defined by some common characteristics. The identification of the population should be followed by the computation of the sample size. The final step would be for the research team to position themselves in vantage positions where they can access the research subjects. They have to pick the right informants till the exhaustion of their sample targets.

In case the research team wants to explore the relationship between university education and income. The research team should hunt for people with undergraduate degree as a minimum qualification. The team should visit strategic places where they can get subjects with degrees. Every element they come across with a degree qualification should be requested to participate in a study. The researchers should continue with the study till the sample size is exhausted. The selection of research subjects is based on availability. The team should gather data from any respondents they meet provided the qualifications for being a responded are met.

The strategy of selecting respondents appears to be simple because it is based on availability. Only samples that are conveniently available will be considered to be part of the study. There may be a challenge of getting competent and proper informants. Not all respondents are competent to provide genuine responses. Some people may fake qualifications they may pretend to be qualified when they are not qualified. It may be very difficult for the research team to verify because some kinds of verification may result in the invasion of privacy of subjects. The question of objectivity may still come to play. A fair selection of research subjects is the one that gives all elements of the population and equal chances of being selected.

10 SELF-SELECTIVE SAMPLING

There are other qualitative studies in which the research subjects volunteer to be part of the study. Self-selective sampling is often called volunteer sampling. The selection of samples is based on volunteering basis (Fink, 2003). The research team should follow some basic steps during the processes of subjects' selection and data gathering.

The first step is for the research team to identify the population. The identification of the population depends on the nature and the context of the study. After the identification of the population, the research team's task will be to convince volunteers to participate in the study. The team should demonstrate that the study is important to them and the community. They should convince the population on why it is important for them to volunteer to participate in the study.

Depending with the nature of the study, it may be prudent for the research team to extend the reciprocity hand to the participants. In case the research team is dealing with sick patients, the team may assist with medicines if the resources are available. The respondents may also be able to gain knowledge about their sickness. The research team should be able to inform the population about the advantages attached to participating to a study as a volunteer.

The major challenge for volunteer sampling is that of convincing elements from the population to volunteer. It is a serious challenge to the studies that have less advantages and benefits to the research subjects. Such kinds of studies are affected by high rates of dropouts if the research subjects are not highly motivated. It is therefore the duty of the research team to keep the levels of morale high. The studies related to qualitative are only concerned with emotions, feelings and experiences of individual respondents and not the quantity. Therefore the challenge of the numbers of respondents to qualitative researchers may not be a problem.

Non-probability sampling procedures have a key role to play in research. They can be adopted in situations where probability sampling procedures cannot apply. It has been noted that probability sampling procedures are not able to access hidden populations. Hidden populations can only be accessed by non-probability sampling procedures like snowball. Furthermore, probability sampling procedures are not able to handle populations with few elements. The duty of sampling procedures is to complement each other. A pragmatic approach of handling sampling of research elements is highly encouraged. The readers of this book should now be able to follow critical steps of selecting respondents via non-probability sampling procedures. There is also need for the researchers to adhere to research ethics because qualitative studies may invade the privacy of individuals if they are not handled with care.

Critical Thinking Challenge

1. Do you think self-selective sampling is credible in qualitative projects? Why

11 QUESTION AND ANSWER SESSION

Why is Snowball Sampling Unique in Qualitative Projects?

Snowball sampling is unique because it is used to identify hidden elements/ respondents. The technique is used to locate elements that are not accessible. The process begins with the selection of one sample/element or few samples/ elements that will lead the researchers to the other hidden samples. It should therefore be clear to readers that snowball sampling procedures is able to locate hidden samples.

What are Hidden Samples?

These are samples/persons not comfortable in giving information in public due to the sensitivity of their situations. There are some research projects that are sensitive like those involved in drugs and corruption. Identifying participant who are drug dealers may be a challenge because they are always in hibernation and the only solution is to use snowball to locate them. There are research projects that are undertaken with HIV patients, the victims are always afraid of stigma and they are always in hibernation. The use of snowball will help to identify such persons.

What are the Steps of Undertaking Snowball Sampling?

The population to be involved in a project is in hibernation and the researcher should begin by identifying a single sample or few samples. The second step is for researchers to get information from the identified sample or few samples and request the sample/few samples to connect them with other samples in hibernation. The researcher should continue doing the same till the target is reached.

What are the Ethical Issues Related to Snowball Sampling?

There are several ethical issues involved. The researchers should guarantee privacy and confidentiality of real identities of the respondents. The process of information gathering should be done in secrecy because the respondents do not want their status to be exposed. They are in hibernation and researchers should follow them in hibernation too. The researchers should consider reciprocity and may give them some gifts or assistance in medical expenses depending with their financial power.

What is Quota Sampling Procedure?

Quota sampling involves the selection of samples from a population based on representation from all strata. The sampling procedure is related to cluster and stratified sampling but the only difference is that quota sampling do not use randomization in any of its stages. The major advantage of quota sampling is that the entire population is represented because the samples are picked from each and every segment of the population.

What are the Types of Quota Sampling?

There are some types of quota sampling. There is proportional quota sampling in which the samples are picked from each quota based on the total population of each segment. The higher the population of a segment, the higher the sample size of that segment. There is also disproportional quota sampling in which equal samples are selected from each and every quota irrespective of their population size. This is done for comparison purpose.

What are the Steps of Undertaking Quota Sampling?

The first step is to identify and study the population of the study. The second step is to divide the population into segments depending with the characteristics of elements. The third step is to compute the quota of each segment via proportional or disproportional representation. The final step is to pick the samples from each segment and get data from the elements till the target is reached.

What is the Purpose of Convenience Sampling?

The key function of convenience sampling is to pick samples or persons to be involved in the study basing on their availability. If the purpose of the study is to interview persons with a first degree programs, the researchers should visit strategic places and interview all people with a first degree programs that they come across with. The selection of samples is done based on convenience and can be used where there is shortage of resources and time to execute the project.

What is it Self-Selective Sampling?

It is sometimes called volunteer sampling. The samples volunteer to become part of the project. The challenge comes when researchers try to convince persons to be volunteers. The research team should come up with some strategies to coerce respondents to be volunteers. They research team may consider employing the reciprocity strategy but the strategy should not be used to spoil respondents. The other challenge is on whether the research team may get key informants because not all volunteers are key informants. This may affect the credibility of the study.

What is Purposive Sampling?

The sampling procedure is often called judgemental, selective or subjective sampling. It is the sampling procedure that is done on purpose. It is mainly connected to case study design. The purposive sampling has several variants and the following variants are discussed in this question and answer session;

- Homogeneous Sampling
- Heterogeneous Sampling
- Typical Case Sampling
- Extreme Case Sampling
- Total Population Sampling
- Expert Sampling

What is Homogeneous Sampling?

It is the type of sampling procedure used to select samples from a population with similar characteristics. The research team should begin by identifying a population with similar characteristics and start to pick samples till the target is reached. There is no need for segmentation because the traits of elements are the same. The research team should make sure that the characteristics of all elements in a population are similar before picking the samples.

What is Heterogeneous Sampling?

The sampling procedure is often called maximum variation. It is the opposite of homogeneous sampling the research team should adopt heterogeneous sampling if the population under the study has elements with different characteristics. A researcher may want to study the attitudes of married people towards marriage. In this case, the researcher may start by identifying a population and use heterogeneous sampling to select samples based on gender and the period or duration in marriage. The purpose of this sampling procedure is to make sure that all the deferent characteristics of a population are represented.

What is Typical Case Sampling?

It is the sampling technique of purposive that is used when the research team is interested in normality or typicality of elements, people, events or cases. The sampling technique is concerned with the usual samples or average samples. For example a researcher may want to examine the live styles of families leaving on average salaries. The first step is to identify the population with elements of families surviving on average salaries and start collecting data from them till the target is reached.

What is Extreme Case Sampling?

The sampling technique is sometimes called deviant case sampling and is the opposite of typical case sampling. Unlike the typical case sampling, extreme case sampling is concerned with the unusual or special cases and events. The purpose of the method is to target population with special or unusual elements and select elements till the target is reached. It is usually good for problem solving because it targets elements that are unusual or deviant from the rest.

What is total Population Sampling?

It is sometimes called comprehensive sampling. It is a unique variant of purposive sampling because all elements in a population will automatically become part of the study. It is used when the population has very few element. In the case of a project with physically handicapped persons in village X the research team may use comprehensive sampling to select all elements of the population if the population has few elements. Imagine a researcher is undertaking a project with Members of Parliament in province B (with only 16 Member of Parliament). All of them may become part of the study.

What is Expert Sampling?

It is the sampling procedure that may be adopted if researchers are dealing with educated and specialised samples. The research team have to select samples from experts only. It is used when there is a shortage of empirical literature. The research team should also consist of experts. A research team may want to study the distributions patterns of Ebola disease in Country Z and they may need to engage experts rather that lay persons.

REFERENCES

Bluman, A. G. (2007). *Elementary Statistics: A Step by Step Approach.* McGraw Hill.

Chaudhuri, A., & Stenger, H. (2005). *Survey Sampling: Theory and Methods.* Taylor & Francis Group, LLC. doi:10.1201/9781420028638

Daniel, J. (2012). *Sampling Essentials: Practical Guidelines for Making Sampling Choices.* Sage Publications. doi:10.4135/9781452272047

Doherty, M. (1994, March) Probability versus Non-Probability Sampling in Sample Surveys. The New Zealand Statistics Review, 21-28.

Fink, A. (1995). *How to Sample in Surveys* (Vol. 6). Sage Publications.

Fink, A. (2003). *How to Sample in Surveys (2nd ed.).* Sage.

Gravetter, F. & Forzano, L. (2012). Selecting Research Participants. *Res. Methods Behav. Sci.*, 125–139.

Madondo, S. M. (2016a). Essentials of Social Science Research (1st ed.). Mount Meru University Publications Unit.

Shorten, A., & Moorley, C. (2014). Selecting the Sample. *Evid. Based. Nurs., 17.* doi:10.1136/eb-2014-101747

Wretman, J. (2010). Reflections on Probability vs. Non-probability Sampling. *Official Stat. Honour Daniel Thorburn,* 29–35.

Zhi, H. L. (2014). *A comparison of Convenience Sampling and Purposive Sampling*. Academic Press.

Chapter 6
The Trustworthiness of Qualitative Research Outputs

ABSTRACT

This chapter exposes readers to the strategies of handling trustworthiness in qualitative research projects. The first section contains information related to the methods of promoting credibility of qualitative research projects and instruments. The computation of coefficient of credibility (CC) is demonstrated in this section. The second section addresses the strategies of promoting transferability of qualitative projects and outputs. There is a discussion on whether the qualitative research outputs and completed projects should be generalized to other contexts. The strategies of enhancing dependability are handled in Section 3, and Section 4 contains the methods of handling conformability in qualitative research projects. The chapter is concluded by a question and answer section.

1 KEY CONCEPTS

- **Trustworthiness** – is the acceptability of qualitative research instruments/outputs, the acceptability is based on credibility, transferability, dependability and conformability.
- **Credibility** – the ability of qualitative instruments to answer the qualitative projects' objectives and the connection between the qualitative outputs and the real situation on the ground.

DOI: 10.4018/978-1-7998-8549-8.ch006

- **Transferability** – the application of qualitative outputs to other similar contexts as well as the replication of qualitative studies to other similar contexts.
- **Conformability** – the control of bias in qualitative projects/have something to do with objectivity.
- **Dependability** – have something to do with the reliability of qualitative projects and outcomes.
- **Table of Specification** – a table with details that should be used in designing qualitative instruments. It can be used to validate the qualitative instruments.
- **Coefficient of Credibility** – a formula used to compute the validity/ credibility of qualitative instruments.

2 LEARNING OUTCOMES

By the end of the chapter, readers should be able to;

- Describe the concepts of credibility, transferability, conformability and dependability in qualitative projects
- Identify strategies for boosting trustworthiness of qualitative instruments and outputs
- Apply trustworthiness strategies in qualitative projects
- Compute the credibility of qualitative instruments via Coefficient of Credibility (CC)

3 CASE STUDY

Mr. Benson was a senior researcher at African Union offices in Addis. He was a team leader of a qualitative project exploring the life styles of indigenous people in Ethiopia. He led a team of six researchers who designed qualitative research instruments within 24 hours. The team agreed that the instruments were to be exposed to pilot testing. The piloting was to be done within a period of five days. After a successful pilot study, the team proceeded to gather data from the research sites. By the end of this chapter, you should be able to note that piloting of research instruments is good for the trustworthiness of qualitative outcomes.

4 INTRODUCTION

What is interesting to note from the case study above is that Mr Benson worked with a team of researchers when designing instruments. The team also exposed the instruments to a pilot study. The aim of all these activities were to improve the trustworthiness of qualitative instruments and outputs. In positivist paradigm (quantitative research), trustworthiness is equivalent to reliability and validity. The positivists modify their research instruments during pilot studies in order to make sure that they are valid and reliable. Silverman (2001) and Pittts (1994) made starling contributions towards the concept of trustworthiness in qualitative study. There are two major issues related to trustworthiness that qualitative research specialists should address. The qualitative researchers should test the credibility of qualitative research instruments. Finally, they should prove that the qualitative research outputs are reliable.

Trustworthiness in qualitative research can be defined as the credibility, transferability, dependability and conformability of qualitative research instruments and outputs. Credibility is equivalent to internal and external validity of quantitative paradigm. Transferability is connected to generalizability, dependability is related to reliability while conformability is equivalent to objectivity when compared with the quantitative paradigm. (Guba, 1981).

The general objective of this chapter is to equip readers and researchers with strategies enabling trustworthiness in qualitative research projects. Trustworthiness can be also defined as the acceptability of research outputs by the consumers of knowledge in the academic market. The research projects that fail to pass the credibility, transferability, dependability and conformability tests will not be acceptable in the academic market. Therefore, this chapter is divided into the following sections;

- Methods of Promoting Credibility in Qualitative Projects
- Methods of Promoting Transferability in Qualitative Projects
- Methods of Promoting Dependability in Qualitative Projects
- Methods of Promoting Conformability in Qualitative Projects

5 METHODS OF PROMOTING CREDIBILITY IN QUALITATIVE PROJECTS

There are two questions that have to be addressed when handling the credibility question in qualitative projects. Are the instruments related to the objectives of the study? The issues related to internal validity should be addressed. The research instruments that are not addressing the objectives or the needs of the study are not credible. Credible research instruments produce credible and accepted research outputs.

What is the connection between the research findings/conclusions and the reality on the ground? Credible research findings and conclusions are the ones that are congruent to the situations on the ground. The issues related to external validity should have to be addressed. The research findings and conclusions that are different from the situation on the ground are said to be not credible. There are several strategies that qualitative researchers should make use of when handling the credibility question.

Use of tables of specification – table of specification should be used by qualitative researchers to design credible research instruments. A credible or valid research instrument is the one that captures all of the objectives of the study. An instrument that is not congruent to the objectives of the study is invalid or not credible. The table of specification should stipulate all objectives to be addressed by the instrument and the number of questions for each objective. For example, a study may consist of four research objectives and the researcher may decide to come up with four questions per each objective. The implication is that the instrument may end up with 16 questions.

It is therefore important for readers and researchers to appreciate that tables of specification help instrument designers to be focused. The tables contain the information about the objectives to be addressed by the instrument and the number of questions to set per each and every objective. Qualitative research instruments that are valid are the ones that are capable of producing credible research outputs.

The use of expert advice – the qualitative researchers may seek advice of experienced qualitative experts when evaluating the credibility of their research instruments. The experts should be able to check the credibility or the internal validity of the research instruments. The experts are expected to check on whether the research instruments and objectives of the project are related. They can do that by computing the Coefficient of Credibility (CC) of each and every qualitative research instruments. In quantitative research, they call it coefficient of validity.

The formula that the qualitative instrument experts should use to test the coefficient of credibility is shown below;

CC = IDV X 100

TNI

CC = Coefficient of Credibility
IDV = Number of Items Declared Valid
TNI = Total Number of Items in an Instrument

The purpose of coefficient of credibility is to compute the percentage of questions in an instrument that are congruent to the research objectives. For example;

A qualitative researcher designed an interview instrument of 17 questions. An analysis by an expert showed that 3 of the question in the instrument were not valid. Compute the coefficient of credibility of the instrument?

{14 (valid questions) divide by 17 (total number of questions in an instrument) multiply by 100 equals to 0.82 (82%)}. The implication is that 82% of the questions in an instrument are credible (related to the study objectives). The remainder (18%) are not credible or valid to the study objectives because they may be questions related to demographic characteristics that are important to the study but not valid to the objectives of the study.

The purpose of computing the coefficient of credibility is to make sure that the research instruments are addressing the objectives or the needs of the study. The duty of the expert advice is to make sure that the qualitative research instruments are credible. Valid research instruments produce credible research outputs. Therefore, the computation of the coefficient of credibility guarantees trustworthiness of qualitative research outputs. The credibility should start with the research instruments and the duty of qualitative researchers is to make sure that their instruments are credible.

Critical Thinking Challenge

1. What is your reaction to the coefficient of credibility formula, does it make sense to you?

Use of appropriate qualitative methods – Yin (1994) suggests that the credibility of qualitative research output is guaranteed by the use of proper/approved methods and instruments in research projects. The question that the readers and the researchers may want to answer is, what are appropriate methods and instruments for qualitative studies? The researchers undertaking qualitative studies are always encouraged to make use of approved methods and instruments.

Qualitative studies start with an appropriate paradigm. The appropriate philosophy or paradigm for qualitative is social constructionist. The qualitative researchers should be always guided by social constructionist philosophy that is highlighted in detail in this book. There are also appropriate designs for qualitative studies. The researchers under taking qualitative studies should select designs from the big five traditions (Creswell & Clark, 2007) and other qualitative designs. Biography, ethnography, phenomenology, case study, grounded theory, historical and other qualitative designs should be used by researchers.

What are the approved sampling procedures for qualitative studies? Qualitative studies make use of non-probability sampling procedure like purposive, quota, convenience, volunteer and snowball. The credibility of qualitative studies is ensured if such kinds of sampling procedures are used in qualitative research projects. The qualitative researchers are always encouraged to stick to the approved sampling techniques.

What are the approved data gathering methods and instruments in qualitative studies? Qualitative research make use of interviews, focus group discussion, observation, documents and other methods of data gathering. The qualitative researchers should make use of open ended instruments i.e. interview guide, focus group guide, observation guide, document analysis guide and others. The qualitative researchers should also make use of appropriate data analysis strategies. They should be able to perform open, axial and selective coding within and outside the grounded theory.

The adoption of well-known and acceptable methods is encouraged. The use of approved qualitative strategies help researchers to produce credible and acceptable research outputs. Testing new methods is qualitative research is discouraged. The researchers should subject such methods for approval processes before they are accepted for use. It is unacceptable for researchers to make use of unapproved methods in qualitative studies.

Randomization in qualitative – randomisation is a concept for quantitative studies but may be used in qualitative studies when selecting research subjects from a population. Randomization may be used in qualitative studies in order to reduce biases and promote credibility (Preece, 1994). The subjective selection of research subjects may compromise the quality of research outputs. The qualitative researchers may borrow randomization technique from quantitative paradigm and use it in qualitative projects to select subjects from populations.

The advantage of randomization is that all elements have equal chances of selection (Bluman, 2007). The data produced out of randomised subjects is deemed to be credible. The pragmatic approach in handling research projects is sometimes good for the quality of the research outputs. Randomization may be used in qualitative studies despite the fact that the technique is typical for quantitative studies. Qualitative researchers are encouraged to borrow qualitative methods that are approved and able to produce credible results.

Critical Thinking Challenge

Should we use randomisation in qualitative projects? Why

Triangulation of methods – triangulation is the use of several methods in research and the strategy is linked to mixed methods research (Johnson & Onwuegbuzie, 2004). However, triangulation of methods can be done in qualitative projects. The question is, which qualitative methods can be triangulated? It is obvious that triangulation approach brings credible results that are detailed (Madondo, 2016a).

Is it possible to triangulate qualitative designs? Yes it is possible qualitative researchers may use two or more qualitative research designs selected from the social constructionists' pool. There is no problem if a qualitative researcher adopts ethnography and grounded theory as designs for a project. The use of research designs is determined by the number of research objective. Different objectives may need different qualitative designs to be executed.

Is it possible to triangulate data gathering methods in qualitative studies? Yes it is common and acceptable in qualitative studies for researchers to adopt several methods in data collection. The number of data collection methods to use may be determined by the types of research designs to be adopted. According to Yin (1994), a case study design is unique because it may adopt several data collection methods in a single project. He suggested six data collection methods that may be used when undertaking a project with a case study design. The use of two or more qualitative methods of data collection in a project can be used for comparison purposes. The researcher may want to make a comparison of findings from deferent data gathering methods. The use of several methods improve the credibility of the research outputs.

Is it possible to triangulate sampling methods in qualitative projects? Yes it is possible. The researchers may make use of several sampling techniques selected from non-probability sampling family (provided it works). If researchers are dealing with population with several segments or population with different elements, they may use different sampling procedures selected from non-probability sampling procedures. Preece (1994) even suggested the use of random sampling in qualitative studies despite the fact that randomization is a strategy for quantitative researchers. The triangulation of sampling procedures is good for the quality research outputs.

Triangulation of qualitative methods is inevitable in qualitative projects. Qualitative researchers are encouraged to triangulate methods and the practice is good for the quality of the research outputs. However, the triangulation should be based on approved and acceptable qualitative research methods. There is no room for trial and error in qualitative studies.

Constant comparative – according to Boeije (2010), constant comparative is the constant and immediate data collection and analysis. It involves the multiple visit of the researcher to the research site. The constant comparative approach is connected to grounded theory but qualitative research scholars are allowed to use constant comparative outside the grounded theory. The processes of constant comparative will enable the researchers to verify their data. The verification of data is good for the quality of research outputs. Researchers are requested to go back to the research sites if there are new issues emerging from the data collected.

Engagement of competent research team – the credibility of qualitative research findings is enhanced by selecting competent researchers to be part of the research team. The selection of competent and qualified people to undertake research projects is good for the quality of the research outputs. A research project that is undertaken by well-known and experienced persons is most likely to be accepted by the research output consumers. There is therefore need for the inclusion of experts in research projects' teams in order to boost the credibility of the research outputs.

In conclusion, the credibility of the qualitative research projects should be the focus of researchers. They should make sure that proper strategies are used to boost the credibility of the research outputs. The first step in enhancing credibility of qualitative projects is to validate the qualitative research instruments. The tables of specification should be used during the designing of research instruments. Experts should be involved in the evaluation of research instruments. They have to compute the coefficients of credibility of all qualitative instruments to be involved. Qualitative researchers should test their instruments during piloting. It is not acceptable to administer instruments to actual subjects before they are verified. Triangulation of qualitative research methods is encouraged because triangulation boosts the credibility of research outputs. The research teams should consist of experts and experienced persons. This is good for the quality of research output.

6 METHODS OF PROMOTING TRANSFERABILITY IN QUALITATIVE PROJECTS

Transferability according to Guba (1981) is equivalent to external validity and generalizability of the quantitative studies. The paradigm war between the positivists and social constructionists was also based on generalizability. The question was, ***to what extend should the researchers generalize/apply outputs of qualitative research projects to other contexts?*** The positivists had no problem with the application of outputs to other contexts because they believe that reality/knowledge is the same everywhere. The qualitative specialist are extremely cautious when it comes to generalization. They believe that contexts are different and the application of studies to other context should only be done to similar contexts.

The post positivists are totally against the application of research studies or outputs to other contexts because contexts are totally different. They believe that researchers should not transfer studies or research output to other context because knowledge/reality is ever-changing. What is happening in Africa is different from what is happening in Europe because contexts are different. The social constructionists (qualitative researchers) support the idea of generalizability but should be done cautiously.

The transferability concept is treated in this section as the applicability of research studies/outputs to other similar contexts. There is also an element of external validity in transferability according to Guba (1981). Transferability can be also viewed as the link between the research conclusions/findings and the situation on the ground.

Strategies for Enhancing Generalizability/Transferability

It is correct to say that the social constructionists (qualitative researchers) support the application of research studies and their outputs to other context. However, the application of such studies to other contexts should be done cautiously. The application should be only done to other similar contexts. There are similarities that should be observed between the two contexts before generalizations are allowed.

Geography and nature of organizations – the qualitative researchers should consider the geography and nature of organizations before applying research outputs to other contexts. The geographical set up of the context in which the study was done should be similar to the geographical context to which application should be done. Different geographical contexts may not be good for the generalizations. The research teams should study the nature of organizations where the research projects were undertaken and the organizations the studies are to be applied.

The organizations should be involved in the same type of business. They should also have the workforce that is almost related. In other words, the organizations should be closely related in terms of their organization culture and business activities. It is in this context that qualitative research projects and their outputs may be applied to other similar contexts. The research projects' reports should supply information related to the geography and the nature of organizations involved in studies. This is done for the purpose of application of such studies to other similar contexts.

In most qualitative reports, there is a section titled 'The Locale or Study Area'. This is the section that the researchers should give detailed information about the geography and nature of the organizations under the study. The section is good for researchers who may want to apply the studies/outputs to other contexts.

Nature of respondents & population – the population characteristics of the two contexts should be closely related. The research team should carefully examine the population of the locations/organizations where the projects were undertaken and the population where applications/generalizations are to be made. The respondents selected from the population should be carefully examined. They have to be examined interns of their characteristics and quantity. This is done in order to make sure that applications/generalizations may go ahead when researchers are sure that the populations and respondents are closely related.

A problem arises when differences in populations and respondents are observed. If such differences arise, application/generalizations of research outputs to such contexts should be avoided. The qualitative researchers allow generalizations of outputs to other contexts provided that contexts are similar.

Methods & instruments used – apart from generalizing or applying research outputs to other similar contexts, qualitative research projects may be replicated to other contexts. Replication is a quantitative terminology that refers to a process of undertaking a completed research project to another context. For example, a study done in Europe about the challenges facing university students may be undertaken in Africa.

The sampling methods of a research project should be clearly stated and may be used by another teams during the replication of qualitative projects to other contexts. The research team should study the sampling techniques carefully before replicating a study to another context. Apart from sampling techniques, the research approaches, designs and data collection methods should be clearly stated as well. The purpose of stating them is to allow the research teams to replicate completed research projects to other contexts. The qualitative philosophy accept the idea of generalization and application of research outputs to other context. However, the contexts should be similar that is why the methods of undertaken a project should be clearly stated. They help researcher to generalise research output to similar contexts as well as replicating certain projects to other contexts.

Strategies of Enhancing Harmony between Research Findings & Reality on the Ground

Apart from generalizability, transferability also involves a component of measuring the external validity of the research findings and conclusions. The research conclusions/findings should be similar to what is happening on the ground (Guba, 1981). A problem arises when the research conclusions/findings are not tallying with what is happening on the ground. There are several strategies that the research teams should make use of in order to harmonize the research conclusions/findings with what is happening on the ground.

The first strategy is for the researchers to verify their data with the subjects on the research sites. The purpose of giving feedback to the participants on the research site is to make sure that they approve the data as a correct picture of what they are experiencing. The research team may make some changes if what they gathered is different from the experiences of the participants on the research sites. The main aim of data verification with subjects on the research sites is to make sure that there is harmony between research conclusions/findings and what is happening on the ground.

The harmony between the research conclusions/findings and what is happening on the ground may also be enhanced by constant comparative. Constant comparative is a process of data gathering and immediate analysis till saturation is reached. The research teams should consider going for a second round of data collection and analysis if there are emerging issues arising. Multiple data collection will enable research projects to produce research outputs that can tally with what is happening on the ground.

The other strategy of harmonizing the research conclusions and findings with what is happening on the ground is for researchers to discuss the research findings with empirical and theoretical literature. The comparison will help them to check on whether their findings are tallying with what is happening on the ground.

7 METHODS OF PROMOTING DEPENDABILITY IN QUALITATIVE PROJECTS

Dependability is equivalent to the reliability of quantitative research projects. There are two key issues related to dependability that should be answered by qualitative researchers. They should check on whether the qualitative research instruments are dependable and reliable. They also need to check on whether the completed research projects can be replicated to other similar contexts. The replication of projects should produce similar conclusions when applied to other similar contexts. A dependable qualitative project is the one that applies to other contexts if replicated. There are several ways in which qualitative researchers can handle the dependability of research instruments and the research outputs.

The researchers should subject the research instruments to pilot testing before they are administered to the research subjects. The pilot testing will be able to identify the instruments' weaknesses and correct them. There are several factors that may affect the dependability of qualitative research instruments like the jargon used, the arrangement of questions and their complexity. The research team should be able to identify this and rectify during the pilot study. A project with instruments that are not reliable will not produce dependable outputs. The projects will not be able to be replicated to other contexts and produce similar results.

The dependability of qualitative research outcomes is also enhanced and tested by using more than two observers on the same phenomenon and test whether the outcomes agree. For example, in a study of the life styles of Hadzabe people in Northern Tanzania, three observers may be send to gather data from the research site at different intervals. The results from the three observers should be analysed for the purpose of identifying their similarities. If there is an agreement from the three observers, such outcome is said to be dependable. If there are some serious differences, other observes may be send to the research sites so that the differences may be rectified.

Triangulation of methods can test and ensure dependability of the qualitative research outcomes as well. Several data collection methods for qualitative should be used to collect information from the same phenomenon. The data collected via interview guide may be compared against the data obtained via an observation guide. The data should agree on findings because the data is from the same population or research site. The agreement of such data means that the research outcome is dependable.

The qualitative researchers may also use constant comparative method to make sure that the research outcome is dependable. The researchers have to visit the research site several times. The purpose of visits will be to collect data again in order to handle the issues emerging from the initial data analysis. The collection of data from the fieldwork more than two times is good for data verification and cleaning. The data that is clean and verified is dependable. The qualitative researchers should consider visiting research sites several times so that their projects will produce dependable results.

The dependability of research instruments may also be influenced by the timing of data gathering. The researchers should select proper timing for visiting research sites. The timing and situations should be conducive for the research subjects to provide reliable data. Wrong timing is likely to result in the collection of information that may not be dependable. For example, a researcher may decide to visit the research site in a rural area during the time when locals are busy in their farms. The researcher may not be able to get enough attention from the research subjects. The researcher should select time that is conducive for key informants to participate in the study.

The adherence to research ethics is also be connected to the dependability of qualitative research projects. There are some ethical issues that promote dependability of qualitative research projects. The research subjects should not be forced to participate in the project. Forcing research subjects to participate in the study may result in the gathering of invalid information that may affect the dependability of the project. The qualitative researchers should follow ethics all the time and the adherence to ethics is good for the dependability of the projects' outcomes.

In a nutshell, dependability of qualitative instruments is enhanced by piloting. The researchers should not avoid pilot studies and the results of piloting will help them to clean their instruments. Reliable and clean instruments will always produce dependable outcomes. The use of constant comparative is also good for providing dependable results. Dependability and transferability are different but closely related because they both handle the issue of reliability of research projects to other contexts. Dependability is declared if similar outcomes are observed. Under transferability, researchers are allowed to replicate projects but to similar contexts.

Critical Thinking Challenge

1. Describe how constant comparative can promote dependability of qualitative projects?

8 METHODS OF PROMOTING CONFORMABILITY IN QUALITATIVE PROJECTS

Conformability refers to the objectivity of the observers/researchers during data gathering and writing of research reports. The qualitative researchers should make sure that their feelings and emotions are not going to dominate in the study. The voice of the research subjects should be heard rather than the researchers/observers' personal feelings. The conformability is a challenge if the researchers are not qualified and experienced to undertake qualitative projects. The qualification/experience of researchers is always positively correlated with the quality of research outcomes. The observer should always work very hard to reduce the investigator bias.

It should be noted that the social constructionists (qualitative researchers) support the interaction between the researcher and the researched. However, the positivists (quantitative researchers) assume that such kind of interaction is a source of bias. The qualitative research specialists are allowed to incorporate their feelings and experiences into the study but should be done accordingly. Their feelings and experiences should not silence the voice of the research subjects. Knowledge in research is generated through the interaction of the researchers and the researched (Creswell, 2014). It is therefore difficult for the researchers' feelings and experiences to be ignored. The researchers should check on how their feelings and experiences should be incorporated into the study.

There are several ways in which the observers may reduce or control the investigator bias. The investigator bias may be controlled by the bracketing of researcher' feelings and experiences (Creswell & Clark, 2007). The researchers should be able to control their personal feelings and experiences about the phenomenon under the study. They should not suppress the voice of the research subjects. However, the bracketing process is a challenge to most observers. Bracketing is more psychological rather than being practical and majority of the observers find it hard to bracket their feelings and experiences. There are no clear steps on how the observers should bracket their feelings and experiences about the phenomenon under the study.

The investigator bias may also be controlled by peer reviewing. The research reports should be reviewed or evaluated by experienced peers. The peers should be able to identify biases and suggest ways of handling them. The use of peer reviewers in in reviewing research reports is important because peers will evaluate reports from neutral grounds. Research reports are not complete if peers have not endorsed them. They are only complete after being approved by the peer reviewers.

The other technique for dealing with investigator bias in qualitative project is team work approach. The studies should be undertaken by project teams. The project teams should consist of experienced and qualified professionals. The team members will be able to regulate the biases of fellow investigators properly. The duties should be delegated according to the capabilities of project team members. The team members should also be able to review the work of their fellows. In other words, team members may also play the roles of reviewers.

Bias in research projects may also be handled by the use of several observers. More than two observers may be used during collection/gathering of data of the same project. Comparisons should be made to check on whether there is agreement of information and findings. The agreement means that the study findings are objective. If there is discord or disagreement, other observers may be send to the research sites to verify the discord.

In quantitative research projects, objectivity is enhanced by the randomization process. The positivists believe that their projects are objective because they make use of randomisation in selecting research subjects from the population. Preece (1994) suggested the use of randomization in qualitative projects. There are instances that the qualitative researchers may use random sampling procedure to select respondents. The use of randomization is important because each and every element is given equal opportunity to be part of the study. The use of randomization reduces investigator dominance in selecting research subjects. The interference of the investigators in selection of respondents from the populations is not good for the conformability of the qualitative projects.

Triangulation of methods is also good for the conformability of the qualitative projects. Triangulation of methods is done for objectivity purpose. The researchers are free to triangulate data gathering methods. They will be able to make comparison on outcomes basing on methods of data gatherings. There should be agreement of data collected by different instruments. Disagreements should be resolved by revisiting the research sites. Agreement among the data gathering methods means that objectivity of the project outputs is not questionable. Triangulation of methods should be done with qualitative tools. The researchers should select methods from the pool of qualitative paradigm. Selection of methods outside qualitative paradigm will make the project to be classified under mixed methods.

In conclusion, the strength of a qualitative research is governed by its credibility, transferability, dependability and conformability. A genuine qualitative project is the one that handles the key tenets of trustworthiness in research projects. The qualitative investigator should come up with appropriate and proper methods of handling trustworthiness in qualitative studies. The methods of trustworthiness should be applicable to practice.

Critical Thinking Challenge

1. State strategies for ensuring conformability in qualitative projects?

9 QUESTION AND ANSWER SESSION

What is Trustworthiness Of Qualitative Studies and Outputs?

Instead of using the terminology trustworthiness, quantitative researchers used terms reliability and validity. Trustworthiness is the equivalence of reliability and validity in qualitative projects. Trustworthiness is the credibility, transferability, dependability and conformability of qualitative instruments and outcomes. Credibility is equivalent to internal validity in quantitative projects, transferability is equivalent to external validity and generalizability, dependability is equivalent to reliability while conformability is equivalent to objectivity. There are critical questions that need to be addressed when handling trustworthiness in qualitative research projects and outcomes.

What are the Critical Questions to be Addressed Under Trustworthiness of Qualitative Studies and Outputs?

- Are qualitative instruments able to measure what they are supposed to measure?
- Are the qualitative projects' findings congruent to what is happening on the ground?
- To what extend should the findings of qualitative projects generalized to other contexts?
- To what extend should researchers replicate qualitative projects/ outcomes to other contexts?
- Are the qualitative instruments and outputs reliable?
- Are the qualitative processes of undertaking projects objective?

What are the Strategies of Improving Trustworthiness?

There are several strategies of improving trustworthiness of qualitative instruments and outputs. The research team should expose the qualitative instruments to a pilot study. They have to test the instruments so as to verify their credibility and internal validity. According to Lincoln & Guba (1985), the research team should have early visits to the research sites. Interactions of researchers and the research subjects is good for the credibility of the study.

The other strategy is to consider using randomization in qualitative projects. Randomization is a concept for quantitative projects but can be used to select respondents in qualitative projects. Randomization is good for reducing bias and handling the challenges of conformability in qualitative studies. The quantitative specialists are always criticising the qualitative projects for being subjective and the use of randomization will handle such challenge.

The other strategy of trustworthiness is that of triangulation of qualitative methods. The research team should use several methods of data collection, sampling and designs selected from qualitative family. The triangulation of qualitative methods will help to improve the credibility of the study and its dependability. Triangulation is also good for verification.

Constant comparative strategy may be used to boost trustworthiness. According to Boeije (2010), constant comparative involves a multiple data collection and immediate data analysis. This is done in order to verify the data and can improve the credibility and the reliability of the qualitative outputs. There is a general consensus among the qualitative researchers that constant comparative can be used outside the grounded theory.

The research team should give feedback to respondents in the research sites for verification purposes. It is always good for the research team to go back to the research sites and verify their data with the respondents. The process is good because it boosts the credibility and the generalizability of the studies to other contexts. The research team should therefore consider verification of data via giving feedback to the respondents.

The use of known and approved qualitative methods in projects is good for the trustworthiness of the projects and outcomes. The qualitative projects that are credible are the ones that make use of known and approved methods of qualitative. They have to make use of proper designs, data collection methods, sampling methods and data analysis methods and avoiding experimenting with new methods.

REFERENCES

Bluman, A. G. (2007). *Elementary Statistics: A Step by Step Approach.* McGraw Hill.

Boeije, H. (2010). *Analysis in Qualitative Research.* Sage Publications Ltd.

Creswell, J. W., & Clark, V. L. P. (2007). *Designing and Conducting Mixed Methods Research.* University of Nebraska-Lincoln, Sage Publications.

Guba, E. G. (1981). Criteria for Assessing the Trustworthiness of Naturalistic Inquiries. *Educational Communication and Technology*, *29*, 75–91.

Johnson, R. B., & Onwuegbuzie, A. J. (2004). Mixed Methods Research: A Research Paradigm Whose Time Has Come. *Educational Researcher*, *33*(7), 14–26.

Madondo, S. M. (2016). *Essentials of Social Science Research* (1st ed.). Mount Meru University Publications Unit.

Merriam, S. B. (1998). *Qualitative Research and Case Study Applications in Education*. Jossey-Bass.

Pitts, J. M. (1994). *Personal Understandings and Mental Models of Information: A Qualitative Study of Factors Associated with the Information-Seeking and use of Adolescents* (PhD Thesis). Florida State University.

Preece, R. (1994). *Starting Research: An Introduction to Academic Research and Dissertation Writing*. London: Pinter.

Silverman, D. (2001). *Interpreting Qualitative Data: Methods for Analysing Talk, Text and Interaction* (2nd ed.). Sage.

Yin, R. K. (1994). *Case Study Research: Design and Methods* (2nd ed.). Sage.

Chapter 7
Introduction to Qualitative Data Analysis

ABSTRACT

This chapter introduces readers to the basics of data analysis and the practical handling of open, axial, and selective coding within and outside the grounded theory. Readers are introduced to segmentation/reassembling, constant comparative, and analytic induction concepts in qualitative data analysis in the first section of this chapter. They should be able to trace the origin of coding of qualitative data in qualitative research. The stages of qualitative data analysis are discussed in the second section. The third section takes readers through the practical steps of open, axial, and selective coding, and detailed examples are given.

1 KEY CONCEPTS

- **Segmentation** – the breaking down of qualitative data into pieces/segments for the purpose of analysis.
- **Reassembling** – restructuring/ordering of pieces/segments into categories or blocks for the purpose of analysis.
- **Qualitative Data Analysis** – the segmentation and reassembling of qualitative data.
- **Transcription** – the writing down of recorded data from the research sites, can be done manually or via technology (computer).
- **Transcriber** – a person transcribing recorded data.

DOI: 10.4018/978-1-7998-8549-8.ch007

- **Recorder** – a person recording data from the research sites.
- **Coding** – the process of breaking up data into pieces/segments and rearranging pieces into blocks/categories/themes.
- **Code** – a label that represents a category/theme
- **Apriori Codes** – codes developed before data gathering and analysis.
- **Inductive Code** – code generated during data collection and analysis.
- **Coder** – a person doing coding.
- **Codebook** – booklet gathering and containing codes.

2 LEARNING OUTCOMES

By the end of the chapter, readers should be able to;

- Appreciate the definition of qualitative data analysis from a practical perspective.
- Appreciate the rationale behind analysis of qualitative data.
- Understand steps of constant comparative in qualitative projects.
- Understand steps of analytic induction.
- Undertake the task of recording data in research sites.
- Transcribe data recorded from the research site
- Appreciate the rationale for qualitative data coding.
- Understand strategies for qualitative data coding.
- Understand the basics of open, axial and selective coding within & outside the grounded theory.
- Perform open, axial and selective coding practically.

3 CASE STUDY

Mr. Kosmas was a team leader for a project related HIV & Aids in Africa. The project team collected data from the research sites. Part of the recorded data got lost when their computer was stolen, the team became confused because the data was not backed. The data available was transcribed by three persons who were hired by the team. After three months, the data related to the HIV patients (including the actual names of respondents) who were part of the study leaked. It was difficult for the team leaders to verify whether the leakage was caused by the research team or the hired transcribers. The project team leader was dragged to the courts of law by the respondents to answer charges related to privacy and confidentiality. By the end of this chapter, readers should be able to acquire skills on how to handle Mr. Kosmas's dilemma.

Critical Thinking Challenge

The challenge of the loss of data and leakage of private information for respondents could have been avoided during the period of data analysis. How?

4 INTRODUCTION

Researchers should expect inconveniences connected to data recording if there is no baking of data. Transcription of data can be a source of data leakage if ethics are not followed properly. This chapter proposes strategies for handling such challenges. The positivists and social constructionists are only able to make decisions about the data after the stage of analysis. The techniques and strategies of analyses vary according to paradigms. The qualitative researchers are concerned with the categorization and reassembling of data while the quantitative specialists are concerned with the quantification of data. Statistical analyses methods assist quantitative researchers to make decisions about their projects. The qualitative researchers are concerned with the thematic interpretation of individual views, emotions and experiences.

5 WHAT IS QUALITATIVE DATA ANALYSIS?

Qualitative data analysis is defined as the conversion to raw data to a form that is appropriate for researchers to make decisions (Kothari & Garg, 2014). It is very difficult for qualitative researchers to make decisions from the raw data. The raw data has to be processed in order to allow researchers to make conclusions. The raw/primary/source/atomic data is unstructured and unrefined. It has to be refined and processed in order for decisions to be made from it.

Bryman (2002) defined qualitative data as the process of ordering data and make sense out of it. The ordering of data is an initial step of handling raw data. The data is arranged in order to enable readers and researchers to make decisions from it. Creswell (2007) supports the same view, he defined qualitative data analysis as a systematic arrangement of data gathered for the purpose of making decisions.

Creswell & Clarke (2011) view qualitative data analysis as the process of dividing and regrouping data into units with the aim of answering research questions. The ordering process involves the categorization of data into units. A precise definition of qualitative data analysis is given by Boeije (2010). She defined qualitative data analysis as the segmentation and reassembling of data. The segments should be represented by codes and the reassembling process should be done according to the rules of qualitative data analysis.

The definitions given above highlights that qualitative data analysis is the processing of converting raw data to a form that enable researchers to make decisions. The first step in processing qualitative data is to segment it in to categories before rearranging blocks/clusters. Readers may be able to understand the practical meaning of segmentation and reassembling after reading through the entire chapter.

Why Analysing Qualitative Data

The qualitative data is analysed in order to convert the raw data to a form that is proper for interpretation. Raw data is unstructured and should be cleaned so that qualitative researchers will be able to read through the data. The raw data is always in a state of disorder. The other purpose of data analysis is to put order in data. The data that is well arranged will enable the readers and researchers to interpret it with minimum challenges.

Analysis of data is done in order for the readers and researchers to make sense out of it (Miles & Huberman, 1994). It is very difficult for readers and researchers to make sense out of raw data and the process of segmenting, ordering and reassembling of data makes it easy for readers to interpret it. It is through the processes of segmentation, ordering and reassembling that data is refined for easy interpretation. Segmentation and reassembling of data are important steps of qualitative data analysis. Chapter nine deals with practical steps of segmenting and reassembling data.

Critical Thinking Challenge

1. Briefly describe reasons for analysing qualitative data?

6 SEGMENTATION OF QUALITATIVE DATA

There are several terms that can be used to define segmentation. According to Boeije (2010), segmentation is the unfolding, unravelling, breaking up, separating, disassembling or fragmentation of data into pieces. This is the first step of qualitative data processing before the data is reassembled. The data is put into pieces/units for the purpose of easy management. The data is segmented according to cases. If there are 12 research subjects (cases) who participated in a project 12 pieces or segments of data should be constructed. Each piece or segment should give details of what each and every subject said.

The pieces/segments of data should be compared. Similarities and dissimilarities between and among pieces should be observed. The researcher should analyse each piece/segment word by word, sentence by sentence or paragraph by paragraph. The segmentation of data is followed by the reassembling strategy.

7 REASSEMBLING OF QUALITATIVE DATA

It is the process of synthesizing, structuring, integrating, putting together, recombining or uniting the segmented data (Boeije, 2010). This is the second step in the management of qualitative data. The segments/pieces/units are put together under related categories. The categories should be represented by codes. Reassembling is a process of restructuring segments/pieces under categories.

The purpose of reassembling is to put pieces or blocks of data together under categories to enable researchers to make comparison of categories. The ordering of data is important because the researchers will be able to make decisions and conclusions about the data. Qualitative analysis starts with the breakup of data into pieces and ends with the reassembling of the pieces into categories.

8 THE CONSTANT COMPARATIVE CONCEPT IN QUALITATIVE DATA ANALYSIS

The concept of constant comparative is central to qualitative data analysis. The concept originated from Glaser and Strauss (Glaser & Strauss, 1967).

The concept was developed to analysis data for grounded theory. According to Leech & Onwuegbuzie (2011) constant comparative is an activity of comparing new data with the previous data. Under constant comparative, the data is collected and analysed immediately. The collection and analysis of data is done on several occasions till saturation is reached.

Every time the data is gathered, the process of analysis begins. New codes are developed in preparation for the next phase of data gathering (Boeije, 2010). The processes of data gathering and immediate analysis may go on till the saturation is reached. Creswell & Clarke (2007) stated that saturation of data is reached if there are no emerging issues. Therefore, constant comparative is done till there is a state of equilibrium between the findings from the analysis and the situation on the ground. Researchers will be able to provide conclusive answers to the questions upon reaching saturation point. Theories are also developed when saturation point is reached.

There are three key issues to understand when applying constant comparative. First, it involves multiple data gathering and several visits to the research sites. Second, the data is analysed immediately after gathering and finally, new codes are developed in preparation for the next round of data gathering and analysis. The new data is compared with the previous data.

There are several advantages of using constant comparative in qualitative research data analysis. It enhances the trustworthiness of qualitative outputs. The credibility of qualitative projects is boosted by the process of constant comparative. Constant comparative makes the research outputs to be dependable. The verification of data is good for the quality of the research outputs.

Steps of Undertaking Constant Comparative

The constant comparative may involve several steps but the following steps are essential. The process starts with the identification of a phenomenon and the first phase of data gathering. The data should be analysed immediately after collection. The initial coding should consist of several categories and the categories should be compared against each other.

In the second phase, new codes should be developed and the codes should be based on the emerging issues from the initial coding. The researchers should visit the research site again for another phase of data collection. The initial categories are supposed to be re-categorised in the second phase of data analysis. The key categories are identified from the initial categories.

New codes may be developed again if there are emerging issues from the latest data analysis.

The third phase after the re-categorization of the initial categories is the identification of a core category. The core code or the popular code should be compared against all codes. The processes of data collection and immediate analysis may going on till there are no new issues emerging. The constant comparative should reach the climax by saturation and the development of a theory.

9 ANALYTIC INDUCTION IN QUALITATIVE STUDIES

Analytic induction is a concept developed from sociology. It is considered be a concept that prompted the development of grounded theory. The steps of constant comparative are closely related to analytic induction. The concept was coined by Florian Znaniecki in 1934 and its origin can be traced to this period (Znaniecki, 1934). There are several scholars who perfected the analytic induction process and among them was Cressey (1953). Cressey is among the founders of the analytic induction. There is a high chance that the constant comparative process was influenced by the analytic induction process if we are to examine the dates of origin of these concepts.

The concept of analytic induction is important in qualitative data analysis because it shaped the way social constructionists analyse their data and it provoked the development of grounded theory and the concept of constant comparative. The following are the stages of analytic induction according to the work of Znaniecki (1934) and Cressey (1953). Analytic induction is defined as a systematic and exhaustive examination of few number of categories till saturation is reached. The analysis will enable qualitative researchers to make generalizations and reliable conclusions.

The first step in analytic induction is the identification of a phenomenon that should be studied. The problem should be identified as well before the observers think of gathering data. Analytic induction is a problem solving approach that has its roots in sociology. Therefore, the observer should begin by identifying a problem that should be solved by the processes of analytic induction.

After the identification of a problem to be examined via analytic induction processes, the observers should formulate hypotheses about the phenomenon. The hypotheses are important because they keep observers focused and they should be tested against the data on the ground or the research sites.

The observers should go for the initial data gathering on the research sites. The purpose of data gathering will be to test the hypotheses against the data on research sites. The gathered data should be analysed immediately after being collected. The outcome should be analysed based on categories. The observers should come up with new hypotheses soon after the analysis. The hypotheses should be crafted based on the emerging issues from the previous analysis. The observers should test the new set of hypotheses on the research sites again. The gathering of data should be followed by its immediate analysis and new sets of hypotheses may be set again and again. The process may be repeated till saturation is reached. Generalizations can be made when saturation is reached and the observers will be able to make conclusions. In grounded theory, the researchers will be able to develop a theory at saturation point.

The analytic induction is not very different from the open, axial and selective coding within and outside the grounded theory. Analytic induction was not initially developed for the grounded theory. It was developed to help sociologists to make exhaustive examinations of phenomena. The proponents of grounded theory took over from there and developed open, axial and selective coding for grounded theory's data analysis. Therefore, other designs (apart from grounded theory) may perform open, axial and selective coding outside the grounded theory in order for researchers to make exhaustive examination of phenomena. The discussion on axial, open and selective coding is discussed in detail in chapter nine. Readers should be able to go through the practical steps of coding qualitative data.

Critical Thinking Challenge

1. Make a comparison of constant comparative and analytic induction?

10 STAGES OF QUALITATIVE DATA ANALYSIS

At this point, readers should be able to appreciate the definitions of qualitative data analysis, the reasons for analysing qualitative data, the segmentation/ reassembling of qualitative data, the stages of constant comparative and analytic induction. The stages of qualitative data analysis are central to the discussion of qualitative data analysis in this book. All these stages of qualitative data analysis are fully explained in this text. The following are the stages of qualitative data analysis that this text discusses;

- Recording of Qualitative Data
- Transcription of Qualitative Data
- Coding of Qualitative Data
- Structure and Writing of Qualitative Reports

11 RECORDING OF QUALITATIVE DATA

There is a general consensus among the qualitative research scholars that the process of data analysis for qualitative projects starts with the recording of qualitative data. However, some may want to believe that the qualitative data analysis begins with the transcription activities. There is no transcription without data recording. Inappropriate recording of data is likely to affect the credibility of qualitative data analysis. Therefore, the qualitative researchers have to be careful when recording their data on research sites. That is the reason for taking recording of qualitative data as an initial step in data analysis in this book.

There are several methods that can be used to record data on research sites but the methods are not limited to the ones discussed in this section. The common and popular method of recording qualitative data is called pen and paper methods. This is a traditional method of recording research data. It has been in use since the time pen and paper were developed. The emergence of technology failed to do away with pen and paper method of recording. The researchers write their observations on paper (diaries or concept maps) during the process of data gathering.

The method may be used in data collection of interview, focus group discussion and observation methods. However, the use of pen and paper method may be a challenge when collecting data from in depth interviews and focus group discussions. Interviewing (becoming an interviewer) and writing at the sometime may pose a challenge to the interviewer. The interviewer may miss some important information from the interviewee while concentrating on writing or may fail to record (write important information) while concentrating on listening. The interviewer may consider bringing interview assistants to take notes only while he/she is focusing on moderating the discussions.

The researcher may consider using tape recorders or audio recorders. They are appropriate for recording voices of the subjects. They will enable the researcher to concentrate on moderating discussions rather than on writing down what the subjects are saying. The researcher will be able to analyse the voice recorders after the sessions of data gathering. The disadvantage of audio recorders is that they won't be able to capture visual data and other kind of data related to non-verbal actions.

The video recorders will come in because they are able to record both verbal and non-verbal data though they present some serious challenges in transcriptions. The researchers may also use cameras to take photographs during data gathering and the strategy is highly connected to observation method of data gathering. The researchers may also collect artefacts from the research site if they are allowed to do so by those in charge of the sites.

There are some ethical considerations for researchers to observe during the recording stage. The researchers should request for permission to use recording instruments before data gathering. The subjects involved have a right to give or not to give consent. It is also advised that the researchers should visit the research sites prior to data gathering. They should make necessary arrangements and backing of instruments in advance. The researchers should protect the information/data from being lost by backing the data with various methods. They may triangulate recording methods as a method of backing the data. The backing of data data/information via technology may be also be a solution. The data may be backed via icloud and other methods.

12 TRANSCRIPTION OF QUALITATIVE DATA

The second step in qualitative data analysis following the recording of data from the research field is transcription. Transcription is the process of writing down data recorded during the data gathering activity. The duty of transcription is to sort the data in order for the purpose of analysis. The data from the recording instruments is refined to a form that makes it possible for researchers to manage it during cording. Transcription is the segmentation of data into pieces or segments based on cases (what the subjects say). The data recorded on research sites should be written down for the purpose of coding. Readers should remember that qualitative data analysis has been defined as the segmentation and reassembling of data. The process of segmentation starts with the transcription of the recorded data. That is why there is a belief that qualitative data analysis starts with transcription activity and not the recording of data.

The persons who are responsible for transcribing data are called the transcribers. Transcribers can be either junior researchers or anyone who underwent transcription training. It is not possible for a lay person to undertake transcription because there are a lot of issues that need to be put under the microscope. There is therefore need for transcribers to get necessary skills before undertaking the task. The transcribers need to be patient because transcribing may be taxing and complex. The transcribers may transcribe more than 1000 pages in several months.

There are two kinds of data that should be captured during transcriptions. Verbal and non-verbal data or both may be captured during transcriptions. The capturing of both verbal and non-verbal data is considered to be detailed, ideal and credible for qualitative studies. Heath (2006) stated that verbal and non-verbal experiences make the communication to be detailed and sound. Transcribing verbal information only may not give a real picture of what is happening on the ground, which is why both visual and verbal data should be transcribed together in order for the results to be detailed and valid. In a nutshell, transcription in qualitative research is the writing down of verbal and non-verbal data obtained through recording for the purpose of preparing for the coding process. The verbal and non-verbal data should be divided into segments represented by cases or subjects involved in the study.

Types of Transcription in Qualitative Projects

There are several types of transcriptions that may be used in transcribing qualitative data. The methods of transcriptions are selected depending with the resources and time available for the execution of the task. This section considers four major common methods of transcriptions that are in use in different fields and qualitative projects. The selection of these types of transcriptions depends upon data collection methods and the preferences of the individuals and organizations connected to the projects. The data can be transcribed from audio, visual, audio visual sources and documents. The following are common types of transcription;

Verbatim Transcription - which is more detailed and is the transcription of everything including verbal and non-verbal information from recordings. The transcribers should be able to capture everything recorded during the data gathering activity. The procedure allows the transcription of repetitions, coughing, laughter and all words and phrases considered to be slang i.e. gonna, wanna, 'cauz etc. However, there are some rules to follow and the rules are discussed in this section. Greatbatch, Heath, Campion & Luff (1995) stated that visual information is vital for data interpretation and they singled out room layout, body orientation, facial expression and gestures as key components of visual information that may be captured.

Verbatim transcriptions are credited for being ideal or genuine but associated with some challenges. They are not only time consuming but also expensive in terms of resource mobilization. The researchers/transcribers opting for verbatim transcription should plan for resources and timing in advance. The project team may consider to make use of a good number of transcribers in order for the transcription project to be completed on time. The researchers or transcribers may opt for intelligent verbatim transcription.

Intelligent Verbatim Transcription (without slang) – the transcriber use common sense to eliminate slang, repetitions and abbreviations are transformed to full grammar and the non-verbal data may not be necessary. However, the transcriber may transcribe the non-verbal data if it is important but the transcriber uses her/his intelligence in order to come up with reliable interpretations. The major advantage of the intelligent verbatim without slang is that the transcribers will be able to deal with irrelevant data during transcriptions. There is no need for transcribers to transcribe irrelevant data. The transcription of irrelevant (verbal and non-verbal) information can be avoided. It is not only a waste of resources for transcribers to code irrelevant data but also a waste of time.

Intelligent Verbatim Transcription (with slang) – the transcriber uses his/her intelligence to come up with genuine interpretation of the interview or audio data, unlike in the above scenario, where slang is captured in original form. The intelligent verbatim transcription without slang is similar to the intelligent verbatim with slang in that they are all efficient in the management of time and resources. They deal with the irrelevant materials during the transcriptions. There is no need for transcribers to transcribe unwanted materials. However, they are all questioned for not being ideal, the purpose of transcription is to expose everything that was captured on research sites. There is therefore need for transcribers to be truthful to readers by exposing them to what they observed and leave the decision of selecting relevant and irrelevant materials to them.

Discourse Analysis - It uses various approaches to data transcribing and it is based on the pragmatic approach of conducting transcriptions. The implication to practice is that the transcribers are allowed to use any methods of transcriptions depending with the need. The transcribers are free to transcribe verbatim transcription (verbal and non-verbal data) if the data is relevant and necessary. In other scenarios, the transcribers may consider using intelligent verbatim transcription if there is need. There is no need for the transcribers to capture verbal and not verbal actions that are irrelevant to the project. The use of the discourse analysis technique should be determined by time and cost of the project as well as the relevance of the transcribed data to the project.

It is true that verbatim transcriptions are ideal and genuine but they are supposed to be supported by intelligent transcription strategies when covering their short falls. The selection of transcription methods should be based on the relevance of data, cost and resources for the project.

Planning for Transcribing

Planning for a project is not an easy task because transcription is time consuming and a dubious process sometimes. Burkey, Jenkins & Highan (2010) stated that, for every hour of recording, a transcriber may take 4 hours of transcribing only verbal data. If one has 40 interviewees, one needs 160 hours of transcribing which can translate to 7 days (of transcribing without sleeping). It is interesting to note that transcribing process is hectic and complex and the project team should take considerable time planning for the execution of such projects. According to Have (1999), it takes 3hrs to transcribe an hour recording of verbal talk only and 10hrs can be taken for transcribing both verbal and non-verbal information. This is a clear indication that transcribing is a time consuming business. There is an agreement among scholars that the process of transcribing is time consuming.

There are various factors that govern the duration of transcribing. The typing skills of the transcriber may shorten or prolong the period of transcription. The project team should look for transcribers who are computer literate and fast in typing. This will enable the transcriptions durations to be accomplished on time. Apart from typing skills, another determinant of the duration of transcription is the quality of recordings. Poor quality recordings may make it hard for transcribers to accomplish their duties on time. They may also lead to the transcription of wrong formation.

The types of transcription to be used also control the duration of transcriptions. Burkey, Jenkins & Highan (2010) and Have (1999) already highlighted that a lot of time is needed for transcribing verbal and non-verbal data (verbatim transcriptions) as compared to the intelligent verbatim transcriptions. Intelligent verbatim transcriptions are said to be good for the effective use of resources and time. There are other characteristics related to the transcriber that shape the time or duration of transcribing. The experience of the transcriber counts. Experienced transcribers are most likely to finish their tasks on time compared to inexperienced transcribers. Accents of interviewees have influence on the duration of transcribing. The transcribers may take long to accomplish their tasks if they encounter accent problems. The project team should consider all these factors when selecting transcribers in order for the projects to be finished on time. Incompetent transcribers may not only delay projects' accomplishments but may also affect negatively the quality of transcription outcomes.

The transcriber should consider giving extra time of subscribing if there is need. Extra time may be given if the recoding involves more than one interviewee. The transcribers may need enough time to handle such kinds of group interviews. The transcription of focus group discussion data may be complicated and transcribers may be given extra time and resources. Extra time may be added in case of poor quality recordings and the transcribers may need enough time to understand well the poor quality recordings. The same should happen if the transcribers are facing challenges of accents of interviewee and very fast speakers.

Critical Thinking Challenge

1. Look for any audio data recorded. Transcribe the data using manual method or computer?

Ethics Related to Transcriptions

Considering ethics is an integral part of transcribing and consent agreement should be signed by the transcriber. The transcriber should swear and sign consent agreement that, he/she shall follow key ethical considerations during the period of transcription. The transcribers should swear that they will keep the information of the research subjects confidential during and after transcriptions. The privacy and confidentiality rules for protecting research subjects should be observed. Transcribers should not be avenues of leaking personal data for the research subjects.

The transcribers should swear that they will remain objective during transcribing. The transcribers should not manipulate the data of respondents in any way. They should always transcribe what the research subjects say. The voices of the transcribers should not be heard in the transcripts. They should remain neutral and objective. The transcribers should also be warned that the materials for transcription should not be moved outside the transcription laboratories. The transcription activities should be only done within the approved areas. This is done in order to protect the privacy of respondents and the leakage of information. It is in this respect that the transcribers are obliged to sign consent agreement forms before being engaged.

Language and Symbols of Transcription

The researchers and transcribers should understand the symbols and language of transcription and below are some of the symbols and languages suggested by Bailey (2008) which he called 'transcription convections':

- (?) talk too obscure to transcribe
- (.) silence less than half a second
- (..) silence less than one second
- (2.8) silence measured in 10^{ths} of a second
- :::: lengthening of a sound
- Becau- cut off intervention of a sound
- he says. emphases
- (left hand on neck) body language
- Hhhhhh audible out breath
- .hhh in-breath
- [overlapping talk begins
-] overlapping talk ends

Other ways of illustrating symbols language

- (.) short term pause
- (…) medium term pause
- (….) long term pause
- (cough) body language
- (laughter) body language
- (loudly spoken) body language

Example 1 transcription of a conversation between husband and wife without non-verbal language

- ***Husband:*** *You won't study PhD; you should make sure that children are taken care of…..*
- ***Wife:*** *Why why aaaal, my Jesus….*

Example 2 transcription of a conversation between husband and wife with both verbal and non-verbal language

- ***Hus:*** *.hhhh (….), you wn't study PhD (coughing), you should make sure children are taken care of*
- ***Wife:*** *(crying), (….) why, (..) why, (….) aaaal my Jesus (crying uncontrollably)*

Example 1 and two are different though they are portraying the same information. In the example 1 there is no non-verbal language while example 2 has both verbal and non-verbal language and it is clearly showing emotions of the wife and her husband.

In conclusion, transcription process is complex and difficult to handle. The project teams are encouraged to make thorough planning in advance. They should assess their budgets, available time for transcriptions, the recruitment of transcribers and the types of transcription techniques to adopt before progressing to the next step. They also have to sign consent agreement forms with transcribers. This is done to protect research subjects from harm and prevent the leakages of data. The project team should select competent transcribers. This is done in order to enhance trustworthiness of qualitative data.

Critical Thinking Challenge

1. Suggest strategies of backing recorded data during data gathering?

11 PRACTICAL CODING OF QUALITATIVE DATA

Introduction

Coding is a third step in qualitative data analysis according to this book. It is at the heart of qualitative data analysis. The qualitative researchers and readers should be able to appreciate from the previous chapter that after the recording of qualitative data from research sites, the recorded data should undergo transcription process. They recorded information should be written down based on cases. The transcription of qualitative data can be verbatim, intelligent verbatim or discourse analysis. The selection of transcription strategies should be based on the availability of time and resources to execute the project. When the transcripts (from transcriptions) are ready the process of coding should begin. Therefore, this section exposes researchers and readers to the practical steps of coding qualitative data.

Coding was initially designed to analyse data for grounded theory. It is claimed to be the oldest technique of processing data. Despite being a technique of grounded theory data analysis, coding is enjoying widespread usage outside the grounded theory. The use of coding (open, axial and selective) outside the grounded theory is supported by Boeije (2010), Patton (2002), Coffey & Atkinson (1996) and Miles & Huberman (1994). The main objective of this section is to equip readers with practical skill of performing open, axial and selective coding of processing qualitative data.

Definition of Key Terms

Coding is defined as the process of creating ideas, concepts and epiphanies from qualitative raw data (Benaquisto, 2008). The qualitative data gathered from the research sites is reduced to segments for the purposes of interpretation. The data is processed in order for researchers to make sense out of it. According to Boeije (2010), qualitative data analysis is the segmentation and reassembling of data. Categories are formulated and the categories are represented by codes. Example Y below is a simple illustration of a coding process of a data gathered from a study on the experiences of university students in learning statistics.

EXAMPLE Y; DATA CODING (Statement by Statement)

Data Gathered on University Students' Experience in Learning Statistics
Categories & Codes
Strategies for Teaching Statistics **(STS)**
Attitudes towards Statistics **(ATS)**
Challenges of Learning Statistics **(CIS)**
Performance in Statistics **(PIS)**
Factors of Shaping Attitudes **(FSA)**

1. Use of technology/computers is effective in teaching inferential statistics (STS)
2. University students with negative attitude towards inferential statistics can still perform better (PIS)
3. Statistics experience influence attitude of students in inferential statistics (FSA)
4. Computers shape attitude of students in inferential statistics (FSA)

5. University students use inferential statistics in their daily lives (ATS)
6. University students understand inferential statistics equations (ATS)
7. University students suffer from the problem of statistics phobia (CIS)
8. University students put a lot of effort in inferential statistics (ATS)
9. University students often make decisions based on inferential statistics (ATS)
10. University students like inferential statistics (ATS)
11. University students' attitude has no impact on their performance (PIS)
12. My teaching of inferential statistics is completely manual (STS)
13. University students have problems related to language of communication when learning inferential statistics (CIS)

A close examination of the data in example Y shows that the data related to the experiences of university students in learning statistics is divided into four categories that are represented by codes that are in brackets. The categories are; Strategies for Teaching Statistics **(STS),** Attitudes towards Statistics **(ATS),** Challenges of Learning Statistics **(CIS),** Performance in Statistics **(PIS)** and Factors of Shaping Attitudes **(FSA)**. That is a simple example demonstrating the process of open coding. The first step is to segment the data (segmentation) and the four categories identified in example Y are typical examples of segments. The segments are represented by codes. The next step is to reassemble the data according to the identified categories. Therefore, coding is defined as the breakup/segmentation and reassembling/restructuring of qualitative data.

Critical Thinking Challenge

1. Demonstrate your understanding of coding?

Types of Codes

A code is defined as a label that represents a category or a segment. A code can be numerals, words, symbols or phrases. Codes are only used for identification and they do not have any other meaning or function. E.g. of example Y, four categories/segments were identified from the data and they are represented by codes in form of words that are in brackets;

- *Strategies for Teaching Statistics **(STS)***
- *Attitudes towards Statistics **(ATS)***

- *Challenges of Learning Statistics* ***(CIS)***
- *Performance in Statistics* ***(PIS)***
- *Factors of Shaping Attitudes* ***(FSA)***

STS, ATS, CIS and **FSA** are examples of codes there are used to label the categories. These codes are in form of words/letters. They are only used for identification purposes. There are several types of codes but they are not limited to the codes defined below.

Apriori codes – they are sometimes called deductive codes. These are codes that are created or developed prior to data gathering and analysis. There are instances when researchers develop their data collection instrument together with code before going for data gathering and such type of codes that are predetermined are called apriori codes. Most of the standardized research instruments have their apriori codes. The researchers may make use of them during the processing of qualitative data. However, there may be some emerging codes during data analysis different from the ones predetermined. The researchers should consider them as well.

Inductive codes – these are codes that emerge during the examination and processing of the qualitative data gathered. Some of these deductive code may emerge during the data gathering process. The researchers should consider them. Coding of qualitative data is sometimes taxing, the researchers are requested to read the gathered data several times and some new codes may arise during the extensive examination of the data. Such emerging codes are the ones called the inductive codes.

Substantive codes – these are the codes that are linked to the constant comparative activity. They are emerging codes that are a result of multiple data gathering and immediate analysis. The researchers are supposed to gather data several times and analyse it immediately. New codes are developed following the comparison of new data with previous data. Such type of code are the ones called substantive codes.

Codebook – the qualitative researchers are encouraged to have codebooks. These are pamphlets or booklets used for gathering and storage of codes by the coders. Coders are the persons involved in the process of coding gathered data. The codebooks should contain apriori codes and emerging codes like inductive and substantive codes should be captured during the period of gathering and analysing data. Codebooks are important for coders because they can be used to keep their memories alive. There is a danger of forgetting important information for data analysis if it is not properly backed.

Category – category means a cluster or block. A cluster consists of units of similar characteristics. The categories are usually designed to identify units of similar characteristics. The categories are supposed to be compared against other categories for the purpose of making decisions in qualitative research. The categories are closely related to themes and they are often used interchangeably though there are some subtle differences between the two concepts.

The Grounded Theory

The purpose of grounded theory is to develop theories. There is a belief that theories are grounded in the data. The theories are developed following extensive successive processes of data gathering and immediate analysis. The analysis of data is based on open, axial and selective coding. It is from grounded theory that the concept of coding was developed. However, the open, axial and selective coding are of late enjoying widespread usage outside the grounded theory. There is a general consensus among qualitative researchers that open, axial and selective coding should be performed outside the grounded theory because they are detailed and exhaustive in analysing qualitative data.

Grounded theory is unique compared to other designs in two ways. It is the only design in qualitative that is used to develop theories. Qualitative studies develop theories while quantitative studies test theories. The qualitative specialists believe that theories are grounded in the data. Grounded theory is the only design that multiple data gathering and immediate analysis of data is done. The data is gathered and analysed till saturation is reached. Saturation is reached when there are no longer emerging issues from the data. That will be the point in which theories are developed.

Critical Thinking Challenge

1. Should qualitative researchers perform open, axial and selective coding outside the grounded theory? Why

Reasons for Coding Qualitative Data

According to Tylor & Gibbs (2010), coding of qualitative data is important for retrieval, organisation, identification and assembling of data. Qualitative data analysis involves the reorganization of raw data and is done for the purpose of data interpretation and making reliable conclusions about the phenomena.

The analysis of qualitative data is done for identification purposes. The data is identified with codes that are representing categories. The codes are only used for identification function and not any other. The readers and the researchers should be able to identify data and make sense out of it. According to the rules of data analysis, the data gathered should be identified clearly for the purposes of interpretations of such data.

Another function of qualitative data analysis is on organization. The raw data gathered should be organised in a way that will enable the readers to make sense out of it. The first step in qualitative data analysis is that of ordering raw data. The ordering activity is achieved by making sure that the data is organised according to categories. Each and every category of qualitative data should represent information of similar characteristics. The qualitative researchers should be able to make a detailed comparison of such categories before making necessary conclusions.

The other purpose of qualitative data is to assemble the data. The first stage of qualitative data is to break it up into units or pieces. The units or pieces will then be assembled into categories. The assembling of data makes it easy for readers and coders to interpret it. That is why qualitative researchers define qualitative data analysis as the segmentation and reassembling of data for the purpose of making decisions.

The qualitative data is coded to enable researchers to develop theories. The grounded theory entails that theories are grounded in the data. The purpose of coding is to make sure that theories are exposed from the data. The coders will begin by performing open coding followed by axial and selective coding. Coding process was initially designed to analyse data related to the grounded theory.

There is no qualitative data analysis without coding. It is therefore mandatory for qualitative researchers to acquire skills of coding qualitative data. They should be able to perform open, axial and selective coding. The qualitative raw data can only be meaningful if it is properly coded.

Strategies of Coding Qualitative

Coding of qualitative data is sometimes extensive and involving. There are two ways of coding data. The coding can be done manually (pencil-paper coding) or can be done technologically (computer assisted programs like

Nvivo) may be used to sort qualitative data. The manual analysis of data is unavoidable in qualitative projects. It has to be done even if technology is to be used. The data should be sort manually into categories and then fed into the computer. The duty of computer qualitative data analysis software is not to process data but only to arrange it for the purpose of interpretations. Unlike the qualitative data analysis software, the quantitative data analysis software do not only arrange data but also process it into meaningful outcomes.

Qualitative data analysis involves extensive reading of texts/transcripts several times before coding. There are several strategies of coding. Coding may be done word by word. The strategy of coding is detailed because coding is done to the last unit. It involves a lot of reading between words in order for coders to make sense out of the data. Coding can be done line by line or sentence by sentence. The coders should read between lines or sentences in order for them to come up with meaningful codes.

The coding may be done paragraph by paragraph. The coders should read between paragraphs before making reliable conclusions on coding. In some cases, coding may be based on cases. The data of all cases should be categorized into segments. It will be easy for readers to get to know how each and every subject responded. Case coding is acceptable but it is not detailed but general. The strategy of coding may involve all such types of coding. In the case of Nvivo, coding is done based in both case (individual research subjects) and nodes. The nodes are the categories or themes that are found in data. The nodes can be divided into sub-node or sub-categories/sub-themes. This is done in order to refine data to small units. Case coding is mandatory when coders are coding manually because they are able to code basing on line by line/word by word/paragraph by paragraph after careful observation of the data from cases.

Open Coding of Qualitative Data

This is the basic and first step in coding process and it is sometimes called initial coding. The process starts with the development of codes that are used to identify/represent categories. The coders should be able to code the data following the rules. There are some steps that coders should follow when performing an open coding. The steps of open coding are explained in this section as well as the practical examples of performing open coding using pencil and paper method.

Practical Steps of Open Coding

The coders should make sure that information from transcriptions/transcripts is ready for the coding to begin. The documents containing transcribed data are called transcripts. The transcribers should be able to transcribe data from the recording instruments. The previous chapter highlights essential steps of handling transcriptions. In a nutshell, coding involves the categorization of data from transcripts. The data from transcripts should be categorised according to the themes/categories/segments for the purposes of comparisons and interpretations.

Initial reading of transcripts – coders are requested to read through all the transcripts before taking any actions. They should understand what the transcripts are all about and the data that is contained in the transcripts. They should also understand and identify all cases involved.

Second reading of transcripts – the second reading of the transcripts should be done with the purpose of identifying themes/categories. The coders should be able to identify themes/categories and record them in their codebooks. In some cases, the coders may go through the transcripts as a group this will help them to minimise errors. Teamwork approach in qualitative data analysis is highly encouraged. The purpose of reading the transcripts for the second time is for the coders to identify themes/categories.

Putting codes on themes/categories – now that themes/categories has been identified, the coders should identify appropriate codes to represent the categories. The codes may be inform of numerals/words/symbols/phrase. The purpose of codes is to identify themes/categories and there is no any other function apart from identification. In case of a study of the experiences of university students in learning statistics in example Z, four themes/categories were identified after a careful examination of data. The theme/categories are represented by the codes in brackets.

Categories & Codes

- *Strategies for Teaching Statistics* ***(STS)***
- *Attitudes towards Statistics* ***(ATS)***
- *Challenges of Learning Statistics* ***(CIS)***
- *Performance in Statistics* ***(PIS)***
- *Factors of Shaping Attitudes* ***(FSA)***

Codes are labels that are used to represent segments/categories. In the case of the above example, the codes are inform of words and they are used to represent the four categories of data. The process of coding of categories may begin if the codes are identified.

Critical Thinking Challenge

1. What is your understanding of categories and codes?

Coding of data – now that the categories and codes are identified, the coders will now go ahead coding. The coders may read transcripts word by word, statement by statement or paragraph by paragraph. They should read through write codes on margins or end of each and every statement or paragraph. Example Z below is an example of coding (statement by statement/line by line) and the codes are put in brackets at the end of every statement.

EXAMPLE Z; DATA CODING (Statement by Statement)
Data Gathered on University Students' Experience in Learning Statistics

1. Use of technology/computers is effective in teaching inferential statistics (STS)
2. University students with negative attitude towards inferential statistics can still perform better (PIS)
3. University students use inferential statistics in their daily lives (ATS)
4. University students understand inferential statistics equations (ATS)
5. University students suffer from the problem of statistics phobia (CIS)
6. University students put a lot of effort in inferential statistics (ATS)
7. University students often make decisions based on inferential statistics (ATS)
8. University students like inferential statistics (ATS)
9. University students have problems related to language of communication when learning inferential statistics (CIS)
10. University students have challenges in making conclusions in hypotheses testing (CIS)
11. University students have challenges in handling Z,F & T-tests (CIS)
12. University students have challenges in handling rejection and non-rejection zones (CIS)
13. University students have challenges in handling levels of significance (CIS)
14. University students have challenges in handling E & F tables (CIS)

15. University students have challenges in handling critical value (CIS)
16. University students finish their assignments on time (ATS)
17. University students enjoy taking inferential statistics courses (ATS)
18. University students can learn inferential statistics (ATS)
19. University students' attitude has no impact on their performance (PIS)
20. University students are interested in using inferential statistics (ATS)
21. University students are interested in understanding inferential statistical information (ATS)
22. University students are interested in learning inferential statistics (ATS)
23. University students are confident in inferential statistics studies (ATS)
24. University students are afraid of inferential statistics (ATS)
25. University policies influence attitude of students in inferential statistics (FSA)
26. Universities have a shortage of qualified inferential statistics tutors (CIS)
27. Universities have a shortage of infrastructure to teach inferential statistics (CIS)
28. The performance of university students in inferential statistics vary according to disciplines (PIS)
29. The overall performance of university students is above average (PIS)
30. Studying inferential statistics through technology improves performance of university students compared to manual approach (PIS)
31. Students' interests shape their attitude towards inferential statistics (FSA)
32. Students' effort shape their attitudes towards inferential statistics (FSA)
33. Students' confidence shape attitude towards inferential statistics (FSA)
34. Statistics experience influence attitude of students in inferential statistics (FSA)
35. Problem based approach effective in teaching inferential statistics (STS)

36. Poor performance of university students in inferential statistics course is because of their negative attitude towards them (PIS)
37. Performance of university students in inferential statistics influenced by their attitudes (PIS)
38. My teaching of inferential statistics is completely manual (STS)
39. Mathematical background influence attitude of students in inferential statistics (FSA)
40. Lecture method is effective in teaching inferential statistics (STS)
41. Inferential statistics is too technical for university students (ATS)
42. Inferential Statistics is a difficult course for university students (ATS)
43. Inferential statistics formulae are easy for university students to understand (ATS)
44. I teach my students about the value of inferential statistics (STS)
45. I minimize fear of statistics in my students before the training (STS)
46. Genetics influence attitude of students in inferential statistics (FSA)
47. Family background influence attitudes of university students in inferential statistics (FSA)
48. Computers shape attitude of students in inferential statistics (FSA)
49. Case based approach is effective in teaching inferential statistics (STS)

All the 49 statements within a transcript are coded according to the categories/themes identified. A team work approach is needed during the coding process. The members will be able to make sure that reliable data is coded according to the rules.

Reassembling of data – now that the identification of codes has been finished, the data should now be reassembled around the categories. This will enable readers to see through the data easily. All statements coded under a category/theme should be grouped under that them. The regrouping/reassembling should be done like what is in example X below.

EXAMPLE X DATA CATEGORIZATION/REASSEMBLING
Data Gathered on University Students' Experience in Learning Statistics

Attitude towards Statistics (ATS)

1) University students like inferential statistics (ATS)
2) University students enjoy taking inferential statistics courses (ATS)
3) University students are afraid of inferential statistics (ATS)

4) University students can learn inferential statistics (ATS)
5) University students understand inferential statistics equations (ATS)
    1. University students are confident in inferential statistics studies (ATS)
6) Inferential statistics formulae are easy for university students to understand (ATS)
7) Inferential Statistics is a difficult course for university students (ATS)
8) Inferential statistics is too technical for university students (ATS)
9) University students are interested in using inferential statistics (ATS)
10) University students are interested in understanding inferential statistical information (ATS)
11) University students are interested in learning inferential statistics (ATS)
12) University students put a lot of effort in inferential statistics (ATS)
13) University students finish their assignments on time (ATS)
14) University students use inferential statistics in their daily lives (ATS)
15) University students often make decisions based on inferential statistics (ATS)

Factors Shaping Attitudes (FSA)

1. Family background influence attitudes of university students in inferential statistics (FSA)
2. Mathematical background influence attitude of students in inferential statistics (FSA)
3. Genetics influence attitude of students in inferential statistics (FSA)
4. Statistics experience influence attitude of students in inferential statistics (FSA)
5. Computers shape attitude of students in inferential statistics (FSA)
6. Students' interests shape their attitude towards inferential statistics (FSA)
7. Students' effort shape their attitudes towards inferential statistics (FSA)
8. Students' confidence shape attitude towards inferential statistics (FSA)
9. University policies influence attitude of students in inferential statistics (FSA)

10. Performance in statistics (PIS)
11. Performance of university students in inferential statistics influenced by their attitudes (PIS)
12. Poor performance of university students in inferential statistics course is because of their negative attitude towards them (PIS)
13. The overall performance of university students is above average (PIS)
14. The performance of university students in inferential statistics vary according to disciplines (PIS)
15. University students with negative attitude towards inferential statistics can still perform better (PIS)
16. University students' attitude has no impact on their performance (PIS)
17. Studying inferential statistics through technology improves performance of university students compared to manual approach (PIS)

Challenges in Statistics (CIS)

1. University students have challenges in handling levels of significance (CIS)
2. University students have challenges in handling Z,F & T-tests (CIS)
3. University students have challenges in handling critical value (CIS)
4. University students have challenges in handling rejection and non-rejection zones (CIS)
5. University students have challenges in handling E & F tables (CIS)
6. University students have challenges in making conclusions in hypotheses testing (CIS)
7. University students suffer from the problem of statistics phobia (CIS)
8. Universities have a shortage of infrastructure to teach inferential statistics (CIS)
9. Universities have a shortage of qualified inferential statistics tutors (CIS)
10. University students have problems related to language of communication when learning inferential statistics (CIS)

Strategies of Teaching Statistics (STS)

1. Use of technology/computers is effective in teaching inferential statistics (STS)

2. Case based approach is effective in teaching inferential statistics (STS)
3. Problem based approach effective in teaching inferential statistics (STS)
4. Lecture method is effective in teaching inferential statistics (STS)
5. My teaching of inferential statistics is completely manual (STS)
6. I teach my students about the value of inferential statistics (STS)
7. I minimize fear of statistics in my students before the training (STS)

The data has now been reassembled under their themes/categories. This is done for the purpose of easy analysis. The research team should make a critical comparison of categories. The categories should be compared against each other in order for meanings to be attached. If a researcher is asked to give details about the experiences of university students in statistics, he/she may be able to explain the experiences based on four themes/categories i.e. Strategies for Teaching Statistics **(STS),** Attitudes towards Statistics **(ATS),** Challenges of Learning Statistics **(CIS),** Performance in Statistics **(PIS),** Factors of Shaping Attitudes **(FSA)**. Depending with the nature of coding, some studies may end at open coding level while others proceed to axial and selective coding.

Axial Coding of Qualitative Data

Axial coding is a second level of qualitative data analysis. According to the dictates of grounded theory design, the outcome of the initial analysis (open coding) should lead to another round of data collection. This should be provoked by the emerging issues and research teams should be able to come up with a set of new codes. Axial coding is the re-categorization of initial categories. In other words, new categories are developed from the previous categories.

Axial coding is the refinement of the open code. The researchers are supposed to gather and immediately analyse qualitative data. The activity may be done on several occasions. The process should be done till emerging issues are no longer coming out.

Selective Coding of Qualitative Data

The coders/research team should select a popular code from the axial codes. Selective coding is the selection of a core category from the axial codes.

The popular category or core category should be compared with all other categories. The collection of data should continue till saturation. When there are no emerging issues, generalizations are made and theories are developed. The core category should be the centre of all other categories. All other categories should be compared with the core category.

Figure 1. Presents a summary of open, axial & selective coding for issues to consider when undertaking ethnographic studies in Maasai & Hadzabe communities in Northern Tanzania
Source: Madondo (2016a)

Open Coding	Axial Coding	Selective Coding
Research Ethics (RE-Q1) • Got written permission from COSTECH and Pale anthropological Field School at Olduvai Gorge • Consent was based on Verbal agreement	Research approval	Validity and reliability of research data
Gatekeeper and Key Informants (GKI-Q1) • Introduced by NGOs working in Hadza • Connected to gatekeepers by ruling party CCM during campaigning • Gatekeepers helped to link informants • Introduction was easy because the researcher was from the same tribe	Avenues for linking gatekeepers & informants	
Validity and Reliability (VR-Q1) • Triangulation of instruments • Visiting research site more than once • Validation of the report findings by participants • Validating by other related studies	Credibility and conformability of the data	
Reciprocity (RP-Q1) • Gave cigarettes • Paying gatekeepers for the service • Tourists gave second hand clothes • Some never reciprocated	Motivation and research tips	
Data collection (DC-Q1) • Note pads used • Cameras used • Tape recorders used	Instruments of research	

The performance of open, axial and selective coding can be done outside the grounded theory. Table 2 below presents the open, axial and selective coding that was done outside the grounded theory in a study with researchers experienced in undertaking qualitative projects within Hadzabe and Masaai tribes in Northern Tanzania. Figure 1 below presents a summary of open, axial and selective coding for issues to consider when undertaking studies in Maasai and Hadzabe communities in Northern Tanzania.

In conclusion, open, axial and selective coding were initially designed for grounded theory. However, there is a general consensus among qualitative scholars that they should be used in analysing data for other qualitative designs. Performing open, axial and selective coding is detailed and exhaustive that is why researchers are advocating for their coding outside the grounded theory. The researchers are encouraged to observe essential steps of qualitative data analysis. They should make sure that all the rules are followed from data recording to data coding. The coders should also acquire skills of transcription of qualitative data. The transcription of qualitative data should be done following proper methods and consent agreement should be signed with the transcribers.

Critical Thinking Challenge

1. From a practical perspective, define qualitative data analysis?
2. Differentiate qualitative data analysis from quantitative data analysis?
3. Why should we perform open, axial and selective coding outside the grounded theory?

REFERENCES

Bailey, J. (2008). First Steps in Qualitative Data Analysis: Transcribing. *Family Practice*, *25*(2), 127–131. doi:10.1093/fampra/cmn003 PMID:18304975

Benaquisto, L. (2008). Codes and Coding. In The Sage Encyclopedia of Qualitative Research Methods, (pp. 86-89). Sage.

Boeije, H. (2010). *Analysis in Qualitative Research*. Sage Publications Ltd.

Bourque, L. B. (2004). Coding. In *The Sage Encyclopedia of Social Science Research Methods* (Vol. 1, pp. 132–136). Sage.

Bryman, A. (2008). Social Research Methods (3rd ed.). New York: Oxford University Press.

Charmaz, K. (2006). *Constructing Grounded Theory*. Sage.

Charmaz, K. (2010). *Constructing Grounded Theory*. Sage.

Cheril, H. (2009). Theoretical Coding in Grounded Theory Methodology. *Canadian Journal of Nursing Research, 3*(8).

Coffey, A., & Atkinson, P. (1996). *Making Sense of Qualitative Data, Complementary Research Strategies*. Sage Publications.

Cressey, D. R. (1953). *Other People's Money*. Free Press.

Creswell, J. W. (2007). Research Design. Qualitative, Quantitative and Mixed Methods Approach (2nd ed.). Thousand Oaks CA: Sage Publication.

Creswell, J. W. (2013). Qualitative Inquiry and Research design (3rd ed.). London: Sage.

Creswell, J. W., & Clark, V. L. P. (2007). *Designing and Conducting Mixed Methods Research*. Sage Publications.

Gibbs, G. R. (2007). *Thematic Coding and Categorizing: Analyzing qualitative Data*. Sage Publication, Ltd. doi:10.4135/9781849208574.n4

Glaser, B. G., & Strauss, A. (1967). *Discovery of Grounded Theory*. Aldine.

Greatbatch, D., Heath, C., Campion, P., & Luff, P. (1995). How to Desk-Top Computers Affect the Doctor Patient Interaction? *Family Practice, 12*(12), 32–36. doi:10.1093/fampra/12.1.32 PMID:7665038

Green, J., Franquiz, M., & Dixon, C. (1997). The Myth of the Objective: Transcribing as a Situated Act. *TESOL Quarterly, 31*(1), 172–176. doi:10.2307/3587984

Have, P. (1999). *Transcribing Talk in Interaction. Doing Conversation Analysis: A Practical Guide*. Sage.

Heath, C. (2006). *Body Work: The Collaborative Production of the Clinical Object. In Communication in Medical Care: Interaction between Primary Care Physicians and Patients* (1st ed.). Cambridge University Press.

Kothari, C. R., & Grarg, G. (2014). *Research Methodology* (3rd ed.). New Age International Publishers.

Madondo, S.M. (2016b). Experiences in Conducting Ethnographic Studies in Marginalized Communities. A Case of Maasai and Hadzabe Tribes in Northern Tanzania. *Imperial Journal of Interdisciplinary Research, 2*(12).

Miles, M. B., & Huberman, A. M. (1994). *Qualitative Data Analysis: An Expanded Sourcebook*. Sage Publications.

Patton, M. Q. (2002). *Qualitative Research & Evaluation Methods*. Sage Publications.

Znaniecki, F. (1934). *Coming up with the Concept Description: The Methods of Sociology*. Farrar & Rinehart.

Chapter 8
The Structure and Writing of Qualitative Reports

ABSTRACT

The first section of this chapter contains information related the structure of the introduction of the qualitative projects' reports. Readers should be able to compile appropriate background to the study, statement of the problem, research objective, research questions, significance of the study, and other requirements for qualitative projects. The second section contains details related to the review of literature. The readers should be able to understand and apply empirical and theoretical literature in qualitative projects. The third section discusses the structure of methodology of qualitative projects and the strategies of executing it. The analysis and discussion of qualitative data is discussed in Section 4. The final section contains the concluding structure of qualitative research reports.

1 KEY CONCEPTS

Structure of Qualitative Reports – are important subheadings of qualitative reports.

Statement of the Problem – is the description of problem in a research project.

Objectives of the Study – what a project should achieve by the end of its execution.

DOI: 10.4018/978-1-7998-8549-8.ch008

Theoretical Literature – the discussion of the study with the related theories.

Empirical Literature – literature based on five senses from journal articles, published and unpublished theses/dissertations.

Research Methodology – the overall plan for qualitative data gathering.

Response Rate – the actual number of respondents who participated in data gathering as compared to the number proposed by the researcher.

Demographic Characteristics – Bio data of respondents.

2 LEARNING OUTCOMES

By the end of the chapter, readers should be able to;

- Identify and address components of the introductory structure of qualitative reports.
- Understand the basic types, steps of undertaking and writing literature review section.
- Identifying and addressing components of methodological structure of qualitative reports.
- Understand the demands of analysis & discussion sections of qualitative projects.
- Address components of concluding sections of qualitative reports.
- Acquire practical skills/knowledge of writing qualitative reports.
- Write ideal qualitative reports with minimum challenges.

3 INTRODUCTION

Qualitative and quantitative reports' differences are based on research philosophy. The reports differ in terms of their structure and content. By now, readers should appreciate the fact that qualitative projects are controlled by the inductive methods while quantitative projects are shaped by deductive projects. The aim of this chapter is to equip readers with practical skills of writing qualitative projects. The readers should be able to handle qualitative reports with minimum challenges. Therefore, this chapter covers the following components of the qualitative reports structure;

- Introductory Structure of Qualitative Reports.

- Literature Review for Qualitative Reports.
- Methodology for Qualitative Reports.
- Analysis & Discussion of Qualitative Data
- Concluding Sections of Qualitative Reports

4 INTRODUCTORY STRUCTURE OF QUALITATIVE REPORTS

In most cases, the chapter one of qualitative reports is called introduction. The introduction consists of several subheadings but they are not limited to the subsections suggested under this section.

- Background to the Study
- Statement of the Problem
- Research Objectives
- Research Questions
- Significance of the Study

Background to the Study

The main purpose of the background to the study section is to describe the problem with the aim of connecting readers to the statement of the problem. The background to the study trace the problem through international, regional, national and local literature. Researchers are supposed to expose the problem from different contexts and finalise with what is happening on the research.

Readers should therefore appreciate that a powerful background to the study is the one that connects the research problem with the statements of the problem. It should be supported by literature from other contexts. According to the dictates of social constructionist philosophy (which gave rise to qualitative paradigm), the researchers are encouraged to discuss empirical literature that is similar or closely related to the context under the study. Qualitative researchers should only generalize research outputs to other similar contexts (Guba, 1981).

The international, regional, national and local empirical literature should be closely connected to the situations and experiences under the study. The researchers are therefore encouraged to make a comparison of the problem under the study with other closely related problems from other contexts. This is done in order to boost the trustworthiness of the study. A study that discusses empirical literature that are not closely connected to the research problem is questionable in terms of its credibility. In a nutshell, the background to the study should give readers a picture of problems on research sites. It has to be connected to the statement of the problem.

Statement of the Problem

Statement of the problem is the description of a crisis or an interesting phenomenon that needs to be solved by the study. The statement of the problem should identify variables to be investigated. The statement of the problem helps the researcher and reader to identify the purpose or the objectives of the study. The statement of the problem should be clearly answered by the research objectives. There should be a direct link between the statement of the problem and the research title. If there is a difference between the statement of the problem and the research title, the credibility of the research may be compromised.

The following are some of the key characteristics of a statement of the problem according to Bwisa (n.d)

- The statement of the problem should expose and address the research gap.
- Should contribute to the existing body of the research. The statement of the problem should be investigated through data collection.
- Should be of interest to the researcher and suits his/her skills, time and resources.
- The statement of the problem should be ethical.

The statement of the problem should be precise and stated in an unambiguous language. The general consensus on the statement of the problem is that it should be started or expressed in few words and can hardly be more than a page in length. When writing the statement of the problem, the following questions should be asked and answered;

- What provoked you to select the research title?
- Is your title a problem or controversial?
- What variables are to be investigated?
- Is your study worth undertaking?

Steps of Writing a Statement of the Problem

Step One: The researcher must describe the goal or the expectations according to the accepted standards that should prevail. The statement should expose what should be done under the normal circumstances.

Example: A study of the causes of infant mortality rate in country X"
Statement one – it is the duty of the government to provide health facilities to the citizens. According to the constitution of country X each and every person has a right to life. Infants constitute 16% of the total population of country X.

Step Two: is the description of the current situation in a county/place that is causing discontent to the people and the statement should highlight the problem or discontent. Statistics should be given and relevant sources should be supplied.

Example: Statement two; despite measures taken by the government X to reduce infant mortality. Evidence on the ground has shown that the mortality rose by 10% in 2018. The situation is not healthy and it is a time bomb to the society. It is a threat to future economic development.

Step Three is the section where the need to conduct a study is highlighted. It is a statement of the justification for conducting a study.

Example: Statement Three; in response to the increase of infant mortality rate in country X the researcher wants to make an investigation in order to expose the causes of infant mortality rate in country X.

It is very interesting to note that the statement of the problem should consist of three statements as shown above.

- The first statement is a description of the expected condition or situation in an area. The description should focus on what should be expected under normal situation.
- The second statement is a description of a problem that causes discontent in an area. Statistics quoted from relevant sources should be provided in order to highlight the problem. The second statement should highlight the real situation on the ground.

- The third and final statement is the justification for carrying out a study in order to investigate the problem highlighted in the second statement. The researcher should expose the need to carry out the research project. The three statements should be short, precise and should be written in a length of not more than a page. The researcher should write few words which are clear.
- The statement of the problem should be supported by quotations from books, internet, newspapers, journals and other sources. This will provide enough evidence to the problem

Another example of the statement of the problem below;

Research Title: "An exploration of the unemployment rate in country Z"
Statement of the problem

The duty of any country in this world is to provide employment to its citizens and the country needs to create environment conducive for investors to invest in country Z, the ruling party of country Z promised in its manifesto to create one million jobs by 2018.Unfortunately, despite the promise made by the government to create one million jobs, unemployment is the order of the day. The standard newspaper of January 2018 estimated the unemployment rate at 72%. According to the projection from the World Bank (January 2018), the economic growth should go down by 2% by April 2018. Companies are retrenching and some are closing. It is because of the high rate of unemployment and retreatment of workers that forced the researcher to make a qualitative analysis of the employment problems in country Z.

Critical Thinking Challenge

1. Select a research problem of your choice and write a statement of a problem in less than half a page?

Objectives of the Study

Every research project has objectives. The function of objectives is to state clearly what the projects intends to accomplish by the end of its execution. The project objectives should be answered by the literature review and the data gathered from the research site. A credible qualitative study is the one in which all the objectives are addressed in qualitative instruments. The computation of coefficient of credibility should be done to check the connection between the project objectives and what is in instruments.

The qualitative research objectives should be SMART. They should be stated in point form.

- S – Specific
- M – Measurable
- A – Achievable
- R – Realistic
- T – Time

Examples of objectives on a study of experiences of university students in learning statistics.

- *Objective 1 – to explore challenges facing university students in learning statistics at African University in Zimbabwe.*
- *Objective 2 – to examine the attitudes of Africa University students in Zimbabwe towards statistics.*
- *Objective 3 – to assess the connection between performance in statistics and learning environment of students at Africa University in Zimbabwe.*

A close examination of the objectives above shows that the project is specific to context i.e. Africa University in Zimbabwe. However, Africa University context is broad. The project could have been narrowed to a specific program or faculty. The objectives should be specific to contexts. These are some of the issues that the students of research should address because the objectives should be narrowed to specific contexts.

Research Questions

The research questions of a project just like objectives should be specific. The qualitative research questions seek to answer the (why, how & what) questions (Hancock, 2002). According to the dictates of social constructionist philosophy, the qualitative questions and instruments should be open ended and the respondents have freedom of expressing their opinions on phenomena in question. Undertaking qualitative projects is said to be extensive because researchers work with open ended instruments that require a close attentions.

The generation or designing of research questions is easy. Research questions of a project are generated from the research objectives. Research questions can be generated by converting the research objectives to question. If there are three research objectives in a project, there should be three research questions as well. The first, second and third objectives should be related to the first, second and third questions respectively. An example below shows the conversion of research objectives to questions. Just like the research objectives, the research questions should be stated in point form as well.

Examples of the research questions on a study of experiences of university students in learning statistics.

- *Question 1 – what are the challenges facing university students in learning statistics at African University in Zimbabwe?*
- *Question 2 – why are Africa University students in Zimbabwe having unfriendly towards statistics?*
- *Question 3 – how is performance in statistics related to the learning environment of students at Africa University in Zimbabwe?*

Critical Thinking Challenge

1. By using the research problem that you selected above, write two study objectives and two study questions?

In quantitative studies, the research questions of a project may be followed by research hypotheses but qualitative studies are not concerned with hypotheses testing. Most of the qualitative proposal and reports do not state hypotheses. Stating hypotheses is the business of quantitative projects. While the qualitative projects (grounded theory in particular) develop theories, quantitative researchers test those theories (Creswell & Clarke, 2007). This section did not discuss hypothesis testing under this chapter because the testing of hypothesis is mainly connected to quantitative projects. This book is largely concerned with qualitative projects. However, this does not mean that hypotheses are not stated in qualitative projects. The processes of analytic induction approve the stating of hypotheses.

Significance of the Study

The section of the significance of the study is important to policy makers, project funding agents and readers. Policy formulation is directly connected to research outputs. The authorities responsible for policy formulation will get advice and direction from the research outputs. Research outputs give policy makers details about what is happening on the ground and they will be able to formulate appropriate policies that are contextual.

The research project funding agents may only want to fund those research projects that are beneficial to communities or that have a significant impact to the majority of people. That is why they want to study the section of the significant of study when they are evaluating proposals. The readers may also want to know the impact of the research project to the society that is why the section of the significance of the study is pertinent to a research projects.

The significance of the projects should be stated clearly in proposal reports or final reports. Some may state them in point form though there is need to paraphrase them. It is therefore mandatory for researchers to predict the possible contributions of their projects to the community when they are documenting their projects.

5 LITERATURE REVIEW FOR QUALITATIVE PROJECTS

Literature review is the analysis and evaluation of information related to the research title or the research problem. The author has freedom to review literature directly related to the problem or in directly related to the problem. There are some basic procedures to follow when writing literature review however, readers should remember that qualitative researchers should only quote literature that is closely related to the problem under investigation. Use of literature that is not closely related to the study my compromise the credibility of the study's outcome. Literature review is contained in chapter two of the research and the major components to be dealt with in this section are;

- Theoretical literature
- Empirical literature
- The procedures of writing literature review

The first step for a student before writing literature review is to understand the writing styles and this section shall only concentrate on the basics of APA system of writing. APA stands for American Psychological Association and the style of APA system is popular among the universities worldwide. Quite a number of institutions outside America are making use of the APA sstyle.

It is also advised that college tutors should teach APA system or any other styles of writing followed by specific institutions. Readers are therefore advised to visit APA website (http://www.apastyle.org). This section gives some academic requirements of writing theses and assignments according to APA 6th edition of 2010 and one should know that the APA versions change from one period to another.

These are some of the popular dictates of APA styles of writing a thesis or research work.

General guidelines

- Margins should be one inch on all sides of the A4 size paper (top, bottom, left or right)
    - The researchers should make use of double spacing when arranging their work; single spacing is not allowed.
    - The APA recommends the use of 'Times New Roman' 12-pt font size and the students are therefore not allowed to use any other font sizes.
    - The paragraphs should be between five (5) to seven (7) lines anything outside 5 – 7 lines is against the APA style of writing.
    - Each and every heading should not exceed 50 characters anything above is against the APA system of writing.
    - The students are encouraged to use active voice rather than passive *e.g. We predicted that ... and not .... It was predicted that ....*
    - Abstract according to APA should be in one paragraph and begins on a new or fresh page the abstract should be cantered and the length of an abstract depends on colleges or journals but usually the abstract should contain between 150 – 250 words. In other words the researcher should follow the requirements of individual institution on the lengths of abstracts.
    - Researchers should be also careful on the use and proper labelling of tables according to APA as well as the labelling of figures.

Figure 1. below shows an example of illustrating and labelling a table following APA style. The heading should be in italics and on top of the table (not below)

Figure 1. Factors shaping university students toward inferential statistics Source: Author

Factor of IS	N	SD	D	NS	A	SA
Genetics	46	34 (74%)	6(13%)	2(4.3%)	2(4.3)	2(4.3%)
Computer	46	10(21.7%)	11(23.9%)	10(21.7%)	5(10.9%)	10(21.7%)
Students' Interest	46	11(23.9%)	9(19.6%)	11(23.9%)	4(8.9%)	11(23.9%)
Students' Effort	46	11(23.9%)	8(17.4%)	10(21.7%)	6(13%)	11(23.9%)
Family Background	46	12(26.1%)	8(17.4%)	10(21.7%)	11(23.9%)	5(10.9%)

Key N - Total Sample **SD** - Strongly Disagree **D** - Disagree **NS** - Not Sure **A** - Agree **AS** - Strongly Agree **IS** - Inferential Statistics Source: Research Data

Figure 2 below shows an example of illustrating and labelling a figure following APA style. The heading should be not in italics and located below the figure (not above)

Figure 2. Conceptual framework
Source: Author

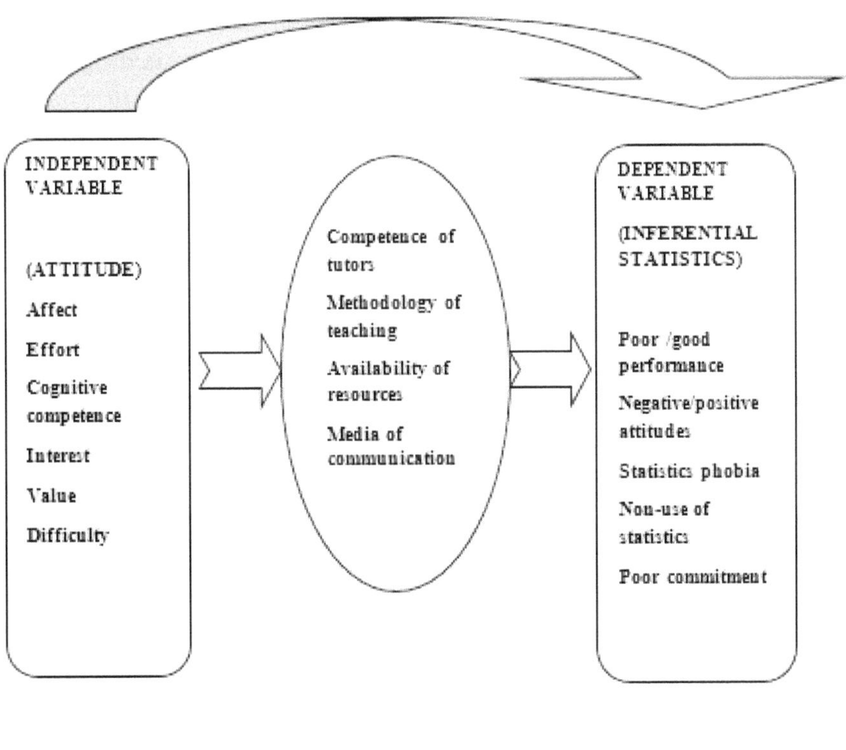

Referencing

This section of referencing is divided into two

- Referencing inside (within) text
- Referencing outside text

Referencing Within Text

- When referencing a textbook or a journal article inside the text of the research project report only a surname of the author and date should be shown example below;

"According to Madondo (2014) ...."

Note that, initials are not required in this section, by initials I mean initials of first names and the middle names are not required.

- In an events that there are six authors or more, the researcher is allowed to write the "Surname" of first author and "et al" meaning and others for example
  According to Madondo et al (2014) ......"
- There are instances were researchers cite personal communication of the author like the emails, letters or telephones conversation which is not recoverable, the researcher should write initials of the first names(s) and surname and the date *e.g. According to the letter or email written by Moyo S.M (January 22, 1999) ...."*

In some cases the date may not be given or missing for example if January 22, 1999 is not known or given the researcher should write (n.d) in place of date for example

According to the letter written or email by "Madondo S.M (n.d) ...."

It is not a crime for dates not to be available and researchers are encouraged to use 'n.d'

- Use only surname and date when quoting from the web e.g.

"Moyo (2016) argued that ...."

You may also cite the entire web for example *"According to the research guidelines of Mount Meru University web http!//*www.mmu.ac.tz ...."

- If one is quoting less than 40 words the statement should be in quotation marks.
- *E.g. "Ethics in* research are controversial and vary according….."
- However if the words are above 40, direct quotation of the statement is not necessary. Summarize the quotation in your own words but remember to acknowledge the author.

Writing Reference Outside Text

- All sources in the reference or the bibliography must be cited in the body. There are instances whereby students write references in the bibliography which are not found in the body. It is an academic crime.
- For textbooks, Authors are cited using Surnames + Initials + Date + Title of the book + Publishing Company+ place of publication e.g.

Madondo, S.M. (2016). *Essentials of Social Science Research.* Mount Meru University Research & Consultancy Unit, Arusha. This is done for a textbook citation.

- For journals, include, Surname + Initials + date + Title + Name of journal + Volume/issue + Page(s) e.g.

Ngatunga, A. (2013). Relevance of Education to Employment. *General Education Journal, Volume 1* (3), 214 – 230.

The researcher should also check on how to reference internet materials according to the APA standards.

It is therefore necessary for the researchers to understand the essentials of APA or any other style because they get information about how to write theses or assignments. APA style is popular among the universities on universe and students are encouraged to study it in detail.

Critical Thinking Challenge

1. Select any 2 textbooks and 2 journal articles from library or online, write outside text reference of the four books and articles following APA standards?

Two Basic Types of Literature Review

Literature review is very important when conducting research because;

- It helps the author to avoid repetition. A lot of researches are being conducted and researchers are encouraged to review literature before the study progresses in order to avoid loss of time and resources because it is not advisable to repeat what has been researched unless unique areas are being observed.
- Literature review will help the researchers to identify the research gap. Each and every research project should open or identify a gap through literature review. If the gap is not identified therefore there is no need for proceeding with the research project.
- We review literature in order to understand the problem in detail. The researchers should be able to master the problem and this may help them to open gaps to concentrate on.
- The review of literature is done for comparison purposes. The researchers should be able to make a comparison between the current study and the previous studies. This is done in order to identify the research gap.
- Literature review is done for the purpose of getting information about the empirical and theoretical literature and this will lead us to the two basic types of literature review.

There are two basic types of literature review which all researchers should know and these are;

- Theoretical literature
- Empirical literature

Theoretical Literature

Theoretical literature is done through the consultation of textbooks/journal articles and the main aim is to get information about the theories related to the problem. There are several theories that are documented about various problems in research and the researcher is encouraged to master those theories in order to get detailed information about the study. The study of theories related to the problem may help the researcher to open the research gap. The researcher should only select and discuss theories closely related to his/her study.

In order to avoid bias, the researcher should consult from several sources of literature and if the authors are able to consult as many books and journal articles as possible, the research gap may be identified without problems. Depending with institutions and the level of study, scholars may be forced to cite a considerable number of sources. This is done in order to promote the credibility of the study's outcomes.

There is a research design in qualitative research called grounded theory and its main aim is to develop new theories. The qualitative researchers believe that theories are grounded in the data. The researchers should be able to gather and immediately analyse the data on several occasions till saturation is reached. When reporting on theoretical literature, researchers should give a summary of;

- The proponent(s) of the theory.
- Brief details on the origin of the theory.
- Key tenets of the theory.
- Advantages and disadvantages of the theory.
- Details on how the theory is related to the study under investigation.

This should be done to each and every theory that is closely related to the study. The key elements of discussion highlighted above should be part of the discussion.

The Empirical Literature

Empirical literature is the literature review based on five senses for example hearing, seeing, touching, testing and feeling. Some may want to call it evidence based on observation. In research the empirical evidence may be based on the studies or researches done on the same area. Journal articles, published and unpublished dissertations/theses are good examples of empirical literature. Before undertaking literature review the researcher should identify previous studies from International, Regional, National, Provincial, District to local level where the study is to be undertaken.

It is therefore interesting to note that the researchers should study the previous researches done from different closely related contexts. This will help the researcher to master in detail the issues surrounding the problem and open a research gap. When making a review of empirical literature, the researcher should study both published and unpublished dissertations, theses and journal articles. When analysing these theses and articles, the following areas should be summarised and assessed;

- The objectives of the study.
- Designs used.
- The methodology used.
- The findings obtained.
- Areas for further research.
- The limitations of the study.

In most cases limitations which the researchers failed to control may become the research gaps. Study the limitations carefully and try to look for the possible solutions to these limitations in the current study. This may be another way of opening a research gap. The researcher is advised to quote recent articles for example citing articles less than 2 years old is an added advantage and use of literature above 5 years old may be outdated depending with the study. However, use of outdated textbooks may not be a problem if the book was written by a classical/primary scholars. Researchers are encouraged to cite classical literature irrespective of their dates of publication.

Relationship between Literature Review and Research Questions

The main aim of literature review is to answer the research questions and therefore there should be a positive relationship between literature review and research questions. If the literature review is not answering the research questions, therefore the literature becomes invalid. The duty of the researchers is to ensure that the literature review is credible.

The researchers should make sure that the research questions are subtitles of the literature review. The main aim of empirical and theoretical literature is to answer these research questions. The researcher should therefore study the research questions before the reviewing of literature. This may helped the study to be focused. This is why a positive correlation is needed between literature review and research questions.

Sources of Literature Review

There are various sources for literature reviews and the basic ones are highlighted in this section like library and other sources. Library materials are vital for providing information needed for literature review. Library is divided into two; e-library and a library for hard copies. In libraries all sorts of materials are found both hard and soft copies. The e-libraries are becoming prominent today and the development of internet technology is making e-libraries a success. It is pointless to carry huge chunk of books from one point to another in this era of internet and technology. Hard copies in the library may be reduced because they are not environmentally friendly.

In libraries we get journals and textbooks which are helpful in writing literature review. Journals contain researches done in various disciplines and the researchers are encouraged to study and read a lot of sources. In some institutions, students are encouraged to cite journals ahead of textbooks because the journals give information about current researches which is classified as empirical. Newspapers and Magazines are also important sources of literature review. Both empirical and theoretical literature are found in Magazines and Newspapers. These newspapers and magazines are available in libraries.

In libraries, students get published and unpublished theses and should take advantage of them when writing literature review. It is compulsory for all students to allow libraries at different institutions to reproduce their theses for library use. There is an agreement signed by the student in a 'release form' allowing the university library to reproduce the thesis for library use only.

There are a lot of free libraries across the globe and some of these libraries are embassy libraries, British Council libraries and United States Information Services. There are libraries for UNICEF, UNESCO, SARIPS and students are always getting free services from these centres.

Archives and Museums are major source of literature. These places contain important information that can be used by students and in some cases little amount of money may be charged but for students it is always free. Court records, commission reports and parliamentary debates are good sources of literature review and students are encouraged to make use of them.

The students may also benefit from Television programs as well as radio programs. Television and radio programs provide the bulky of empirical evidence and very useful to students and scholars. There are several programs for education and entertainment. Internet is becoming popular today because there are a lot of online journals, e-books, newspapers and magazines which are very useful. Students are now making use of internet than before. The main advantage of the internet is that of providing latest information and helps the researchers to address global and international issues.

Guidelines for Writing Literature Review

Several steps should be taken before and during writing literature review. After studying two authors namely Gulvan (2006) and Rallis (2006), the author managed to condense the procedures or steps of writing literature review into 4 (four) stages. Rallis (2006) come up with seven stages of writing literature review after reading the work of Galvan (2006). My reading of the two prominent scholars gave rise to the following steps of writing literature review.

Step one - Review of APA or any other guidelines of writing essays and theses. The America Psychological Association came up with some guidelines of writing theses which are being employed by several universities worldwide. It is therefore mandatory for students to study the APA system or any other style of writing depending with the requirements of their institutions.

Step two - Select the research title or the research problem. It is a critical stage in research projects, the researchers should be provoked and unconscious selection of research title or problem is not acceptable. The selection of research titles should be related to career development of the researchers.

Step three - It is the preparation and planning of writing literature review. The preparation and planning is divided into two pre-planning and post – planning. Pre planning is done when the researcher is making a plan on the content to be covered on the problem. The plan also includes the gathering of the textbooks, journal articles and other resources needed during the review of literature. The researcher should make sure the relevant literature sources are available before the work starts. The articles published and unpublished should be available.

Post planning is a stage that follows when all the plans and materials are available. The researcher should start reading and summarizing the information in tables each table should contain keywords, useful quotes, research gaps, and summary of findings. The tables may also be called concept map format.

Step four - This is the last step of compiling the material for the literature review and the researcher should make use of the concept map formats when writing the literature report. The rules on APA style of writing should be observed. The stage four is the stage of writing a final review. The main aim is to open the research gaps and to get detailed information about the problems.

6 METHODOLOGY OF QUALITATIVE REPORTS

The discussion of the methodology of qualitative projects is done under the following headings;

- Research Approach
- Research Designs
- Population & Sampling
- Data Collection Methods & Instruments
- Trustworthiness
- Ethical Issues

Research Approach

Research approach is a key aspect to consider in any research projects. The researchers should state and justify the selection of the research approach to be adopted. There are three popular approaches in research studies. These are positivist, social constructionist and pragmatist approaches. The first step in handling research methodology is for the researcher to identify the appropriate philosophy to adopt. The philosophy is important because it determines the designs, sample sizes, data collection tools and data analysis methods to use.

Social constructionist approach is the appropriate paradigm to adopt for all qualitative projects. The approach is connected to qualitative research (Creswell, 2014). The adoption of a social constructionist approach means that the researchers should make use of the accepted and popular procedures of undertaking qualitative studies. In research proposals, researchers are supposed to justify the selection of the approach. The positivist approach is not able to handle the (why, how & what) questions (Hancock, 2002). That is why the social constructionist approach may be used because it can handle such questions well. The justification and description may be done in a minimum length of one paragraph.

Research Design (s)

There are various research designs that are appropriate for different approaches. The designs for pragmatist approach are different to the designs for positivist approach as well as the social constructionist approach. A single project may use more than one designs depending with nature of the objectives and the characteristics of the population. A design is an overall plan for data collection and researchers should be careful when selecting research designs to adopt.

There are several designs for qualitative. Creswell & Clarke (2007) talk about five key designs of qualitative which they called the qualitative traditions of the big five. These are biography, phenomenology, ethnography, grounded theory and case study. There are also other designs outside the big five like historical design and others. The implication is that, the researchers should justify the selection of designs in their proposals. The designs should be appropriate to contexts and the dictates of the projects.

Critical Thinking Challenge

1. John has been asked by his organization to undertake a study on the life style of Masaai people in Northern Tanzania. State and justify one research design that John may adopt?

Population and Sampling

The readers should now appreciate the fact that qualitative projects are not concerned with larger sample sizes but the feelings of individuals. The conclusions of qualitative reports are not arrived at because the majority supported a position. They are only concerned with describing and narrating what individuals say. A population is defined as the total number of elements within the research site. The sample size is selected from the population.

In qualitative projects, non-probability sampling procedures should be used to select elements from the population. Among the non-probability sampling procedures to be adopted are purposive, quota, comprehensive, convenience and volunteer sampling (Wretman, 2010). A lot of questions were raised on the objectivity of non-probability sampling procedures because of the interference of researchers in sample size selection. However, researchers are obliged to select sampling procedures from non-probability family when planning for a qualitative project. The sample size selection depends on the availability of time and resources but the concern is not on huge numbers. The sampling procedures to adopt depends on the characteristics of the population elements and the contexts of research sites.

Data Collection Methods and Instruments

There are data collection methods and instruments that are appropriate for qualitative projects. The selection of such instruments and methods should be justified in the proposals. The qualitative researchers may select the data collection methods in qualitative pool of interview, focus group discussion, observation and document review methods of gathering data. The researchers should note that the quantitative instruments should be largely open ended. The qualitative philosophy dictates that qualitative instruments should be open ended.

There are several instruments for qualitative projects like interview guide, focus group discussion guide, observation guide and document analysis guide. These instruments should be largely open ended. The duty of qualitative researchers should be to make sure that the information gathered by these instruments is properly recorded for the purposes of analysis. All in all, qualitative researchers should adopt open ended instruments.

Trustworthiness

In quantitative studies, the equivalence of trustworthiness is reliability and validity. The qualitative researchers have their ways of testing reliability and validity. Trustworthiness is the credibility, conformability, dependability and transferability of qualitative studies and outcomes.

The researchers should justify the strategies that they will use to ensure that the qualitative studies and outcomes are credible, transferable, confirmable and dependable. The methods are discusses in detail in chapter 7 of this book. I hope that readers are by now able to implement practically the methods and strategies of trustworthiness?

Critical Thinking Challenge

1. Suggest strategies of promoting trustworthiness in qualitative project?

Ethical Issues

This section is critical in qualitative projects. The researchers should state and justify the ethical issues that they have to handle from problem selection through data collection to report writing. Ignorance of ethical rules in qualitative research is sometimes dangerous to researchers and respondents. The discussion of ethics is done in detail in chapter one of this books. I hope that by now researchers are well equipped with practical skills on ethics

Critical Thinking Challenge

State and justify the ethical issues to consider when undertaking a project in ethnic areas?

7 ANALYSIS AND DISCUSSION OF QUALITATIVE PROJECTS

The contents of analysis and discussion are debated under the following headings;

- Response Rate
- Demographic Characteristics
- Analysis & Interpretation of Data Question by Question
- Discussion of Theories and Empirical Literature with Study Findings

Analysis and discussion chapter in qualitative studies is a section where the data is analysed and interpreted by the researchers. It may be discussed under the heading suggested above though not limited to the suggested headings. There should be a section of response rate that shows the number of respondents expected to participate in the study and the actual number of the respondents who participated. This should be done on each and every instrument as well.

The demographic characteristics should be also be discussed in detail. The demographic characteristics should show the bio data of the respondents. After highlighting the details of demographic characteristics, the analysis and interpretation of data should be done question by question. The research questions or the research objectives should become subheadings. The readers want to know the responses of participants under each and every question/ objective of the study.

After the analysis and interpretation of data all study questions, the study findings should be discussed with the theories and empirical literature. The discussion is important for verification purposes. It is always good to understand the connections between the study findings and theoretical/ empirical literature.

8 CONCLUSION SECTION OF QUALITATIVE REPORTS

The section on conclusion is discussed under the following headings;

- Summary
- Conclusion
- Recommendations

This is usually the last chapter of the research reports. The researcher should give a summary of the entire project and key finding of the study. The summary of the project should touch on purpose of the study, approaches, designs, sample size, data analysis techniques and summary of the key findings.

The report should then make necessary study conclusions after a careful examinations of findings. Depending with the number of research objectives or questions, there may be one conclusion or several conclusions. The conclusions should be linked to the research objectives. The conclusion should be an overall assessment of the study.

The last section will be on recommendations. The recommendations should be done after a careful examination of the organizations under the study or research sites. It is wrong to give recommendations that are already implemented by the organisations under the study or recommendations that are currently under implementation. The recommendations should be new and such research organisations may implement them right away.

REFERENCES

Boeije, H. (2010). *Analysis in Qualitative Research*. Sage Publications Ltd.

Borg, W. R., & Gall, M. D. (1989). *Educational Research: An Introduction*. Longman.

Bryman, A. (2008). Social Research Methods (3rd ed.). New York: Oxford University Press.

Cohen, L., & Manion, L. (1994). *Research Methods in Education* (4th ed.). Routledge.

Creswell, J. W. (2007). *Research Design. Qualitative, Quantitative and Mixed Methods Approach* (2nd ed.). Sage Publication.

Creswell, J. W. (2013). Qualitative Inquiry and Research design (3rd ed.). London: Sage.

Creswell, J. W. (2014). *Research Designs: Qualitative, Quantitative & Mixed Methods Approaches*. SAGE Publications, Inc.

Creswell, J. W., & Clark, V. L. P. (2007). *Designing and Conducting Mixed Methods Research*. Sage Publications.

Crotty, M. (1998). *The Foundations of Social Research: Meaning and Perspective in the Research Process*. Sage.

Locke, L. F., Silverman, S. J., & Spirduso, W. W. (2010). *Reading and Understanding Research* (3rd ed.). Sage Publication.

Madondo, S. M. (2016a). *Essentials of Social Science Research* (1st ed.). Mount Meru University Publications Unit.

Marshall, C., & Rossman, G. B. (2010). *Designing Qualitative Research* (5th ed.). Sage Publication.

Merriam, S. B. (1998). *Qualitative Research and Case Study Applications in Education*. Jossey-Bass.

O'Leary, Z. (2014). *The Essential Guide to doing your Research Project* (2nd ed.). Academic Press.

Patton, M. Q. (1990). *Qualitative Evaluation and Research Methods* (2nd ed.). Sage.

Related Readings

To continue IGI Global's long-standing tradition of advancing innovation through emerging research, please find below a compiled list of recommended IGI Global book chapters and journal articles in the areas of qualitative research techniques/concepts. These related readings will provide additional information and guidance to further enrich your knowledge and assist you with your own research.

Abdulla, Y. (2020). Ethical Consumerism in Financial Institutions. In E. A. Al-A'ali & M. Masmoudi (Eds.), *Ethical Consumerism and Comparative Studies Across Different Cultures* (pp. 23–38). IGI Global. doi:10.4018/978-1-7998-0272-3.ch003

Abel, T., & Satterfield, D. (2020). HCI and User Experience Design Education. In R. Mendoza-González, H. Luna-García, & A. Mendoza-González (Eds.), *UXD and UCD Approaches for Accessible Education* (pp. 80–95). IGI Global. doi:10.4018/978-1-7998-2325-4.ch005

Abrahamsson, P., Botterweck, G., Ghanbari, H., Jaatun, M. G., Kettunen, P., Mikkonen, T. J., Mjeda, A., Münch, J., Duc, A. N., Russo, B., & Wang, X. (2020). Towards a Secure DevOps Approach for Cyber-Physical Systems. *International Journal of Systems and Software Security and Protection*, *11*(2), 38–57. doi:10.4018/IJSSSP.2020070103

Adam, I. O. (2020). Using a Qualitative Interpretive Approach in Educational Technology Implementation. In R. Boateng (Ed.), *Handbook of Research on Managing Information Systems in Developing Economies* (pp. 318–331). IGI Global. doi:10.4018/978-1-7998-2610-1.ch015

Adetimirin, A. (2020). Technology Adoption for Teaching. In L. A. Tomei & D. D. Carbonara (Eds.), *Handbook of Research on Diverse Teaching Strategies for the Technology-Rich Classroom* (pp. 70–78). IGI Global. doi:10.4018/978-1-7998-0238-9.ch005

Aerila, J., Lähteelä, J., Mäkelä, T., & Kauppinen, M. (2020). Enhancing the Readership of Teacher Students With Digital Book Subscription Services. *International Journal of Digital Literacy and Digital Competence*, *11*(2), 12–27. doi:10.4018/IJDLDC.2020040102

Ahamed, B. B. (2020). Computational Intelligence in the Reversible Data-Hiding Processes Using Radon Transform. In J. J. Thomas, U. Fiore, G. P. Lechuga, V. Kharchenko, & P. Vasant (Eds.), *Handbook of Research on Smart Technology Models for Business and Industry* (pp. 414–424). IGI Global. doi:10.4018/978-1-7998-3645-2.ch018

Akar, S., & Akar, E. (2020). Is it a New Tulip Mania Age? *Journal of Information Technology Research*, *13*(1), 44–67. doi:10.4018/JITR.2020010104

Akca, M. (2020). An Investigation of the Relationship Between Favoritism and Workplace Deviance. In N. Sharma (Ed.), *Analyzing Workplace Deviance in Modern Organizations* (pp. 114–138). IGI Global. doi:10.4018/978-1-5225-9996-8.ch007

Al-A'ali, E. A. (2020). The Role of Auditing Firms in Promoting Ethical Consumerism. In E. A. Al-A'ali & M. Masmoudi (Eds.), *Ethical Consumerism and Comparative Studies Across Different Cultures* (pp. 39–52). IGI Global. doi:10.4018/978-1-7998-0272-3.ch004

Al-Alawi, A. I., Mehrotra, A. A., Elias, H., Safdar, H. S., & Al-Bassam, S. A. (2020). The Implications of Unethical and Illegal Behavior in the World of E-Commerce. In E. A. Al-A'ali & M. Masmoudi (Eds.), *Ethical Consumerism and Comparative Studies Across Different Cultures* (pp. 152–230). IGI Global. doi:10.4018/978-1-7998-0272-3.ch009

Al Amoush, A. B., & Sandhu, K. (2020). Jordan Learning Analytics Management Systems Innovation (Jordanian Universities). *International Journal of Innovation in the Digital Economy*, *11*(1), 44–59. doi:10.4018/IJIDE.2020010103

Al-Jabri, H., & Al-Busaidi, K. A. (2020). Enriching SME Learning and Innovation Through Inter-Organizational Knowledge Transfer. *International Journal of Knowledge Management, 16*(3), 17–32. doi:10.4018/IJKM.2020070102

Al Kubaisi, M., & Naser, N. A. (2021). Identifying Factors That Influence the Use of E-Wallets and Its Continuance. In Y. A. Albastaki, A. Razzaque, & A. M. Sarea (Eds.), *Innovative Strategies for Implementing FinTech in Banking* (pp. 263–290). IGI Global. doi:10.4018/978-1-7998-3257-7.ch016

Alazali, R. M. (2020). Anti-Money Laundering Practices and Ethical Consumerism. In E. A. Al-A'ali & M. Masmoudi (Eds.), *Ethical Consumerism and Comparative Studies Across Different Cultures* (pp. 10–22). IGI Global. doi:10.4018/978-1-7998-0272-3.ch002

Alenizi, A. S. (2020). A Systematic Literature Review for Understanding the Antecedents of the Digital Open Government Matrix. *International Journal of Electronic Government Research, 16*(1), 1–17. doi:10.4018/IJEGR.2020010101

Alexiou, G., Meimaris, M., Papastefanatos, G., & Anagnostopoulos, I. (2020). LinkZoo. *International Journal on Semantic Web and Information Systems, 16*(3), 1–19. doi:10.4018/IJSWIS.2020070101

Alharbi, J. R., & Alhalabi, W. S. (2020). Hybrid Approach for Sentiment Analysis of Twitter Posts Using a Dictionary-based Approach and Fuzzy Logic Methods. *International Journal on Semantic Web and Information Systems, 16*(1), 116–145. doi:10.4018/IJSWIS.2020010106

AlKhaldi, W. A., & Sarfraz, M. (2020). A Technological Solution for Communication With Deaf Students in the Institutions of Kuwait. In M. Sarfraz (Ed.), *Innovative Perspectives on Interactive Communication Systems and Technologies* (pp. 140–156). IGI Global. doi:10.4018/978-1-7998-3355-0.ch006

Allen-Handy, A., Ifill, V., Schaar, R. Y., Rogers, M., & Woodard, M. (2020). Black Girls STEAMing Through Dance. In K. Thomas & D. Huffman (Eds.), *Challenges and Opportunities for Transforming From STEM to STEAM Education* (pp. 198–219). IGI Global. doi:10.4018/978-1-7998-2517-3.ch008

Almeida, R., Gonçalves, P. A., Percheiro, I., Mira da Silva, M., & Pardo, C. (2020). Integrating COBIT 5 PAM and TIPA for ITIL Using an Ontology Matching System. *International Journal of Human Capital and Information Technology Professionals*, *11*(3), 74–93. doi:10.4018/IJHCITP.2020070105

AlSaffar, G. (2020). Ethical Programs for Patients in Bahrain. In E. A. Al-A'ali & M. Masmoudi (Eds.), *Ethical Consumerism and Comparative Studies Across Different Cultures* (pp. 53–73). IGI Global. doi:10.4018/978-1-7998-0272-3.ch005

Altan, Z. (2020). The Role of Science, Technology, and the Individual on the Way of Software Systems Since 1968. In Z. Altan (Ed.), *Applications and Approaches to Object-Oriented Software Design* (pp. 1–33). IGI Global. doi:10.4018/978-1-7998-2142-7.ch001

Amah, O. E. (2020). Handling Variation in Work Ethics and Values Across Generations in Nigeria. In B. Christiansen (Ed.), *Global Applications of Multigenerational Management and Leadership in the Transcultural Era* (pp. 231–262). IGI Global. doi:10.4018/978-1-5225-9906-7.ch008

Amin, A., Valverde, R., & Talla, M. (2020). Risk Management via Digital Dashboards in Statistics Data Centers. *International Journal of Information Technologies and Systems Approach*, *13*(1), 27–45. doi:10.4018/IJITSA.2020010102

Anim-Yeboah, S., Boateng, R., Odoom, R., & Kolog, E. A. (2020). Digital Transformation Process and the Capability and Capacity Implications for Small and Medium Enterprises. *International Journal of E-Entrepreneurship and Innovation*, *10*(2), 26–44. doi:10.4018/IJEEI.2020070102

Apel Campo, T. (2020). When Ethics Meets Technology. In S. B. Buckley (Ed.), *Promoting Inclusive Growth in the Fourth Industrial Revolution* (pp. 270–286). IGI Global. doi:10.4018/978-1-7998-4882-0.ch011

Arcos-Medina, G., & Mauricio, D. (2020). The Influence of the Application of Agile Practices in Software Quality Based on ISO/IEC 25010 Standard. *International Journal of Information Technologies and Systems Approach*, *13*(2), 27–53. doi:10.4018/IJITSA.2020070102

Arora, A. (2020). Automatic Ontology Construction. In M. Sarfraz (Ed.), *Critical Approaches to Information Retrieval Research* (pp. 109–149). IGI Global. doi:10.4018/978-1-7998-1021-6.ch007

Arora, A. S., & Arora, A. (2020). The Race Between Cognitive and Artificial Intelligence. *International Journal of Intelligent Information Technologies, 16*(1), 1–16. doi:10.4018/IJIIT.2020010101

Asabere, N. Y., Agyiri, J., Acakpovi, A., Nachanja, A., & Awuku, P. (2020). Improving Education Delivery in a Technical University in Ghana Through Mobile Learning Technology. *International Journal of ICT Research in Africa and the Middle East, 9*(2), 35–59. doi:10.4018/IJICTRAME.2020070103

Aytan, Y. S., & Sayan, İ. (2020). Ethics in Management. In U. Akkucuk (Ed.), *Handbook of Research on Sustainable Supply Chain Management for the Global Economy* (pp. 299–309). IGI Global. doi:10.4018/978-1-7998-4601-7.ch016

Azizsafaei, M., & Dadd, D. (2020). Sustainable Supply Chain Management in Iranian Manufacturing Companies. *International Journal of Strategic Engineering, 3*(2), 37–58. doi:10.4018/IJoSE.2020070103

Babb, A. P. (2020). Geometry for Computer Graphics in K-12 Education. In J. Keengwe & P. Wachira (Eds.), *Handbook of Research on Integrating Computer Science and Computational Thinking in K-12 Education* (pp. 45–68). IGI Global. doi:10.4018/978-1-7998-1479-5.ch004

Baeva, L. V. (2020). Social and Existential Threats to Personal Security in Virtual Communities. *International Journal of Technoethics, 11*(1), 1–16. doi:10.4018/IJT.2020010101

Baisu, Z., Mengyi, Q., Xiaoling, J., & Lixin, W. (2020). What Are the Experiences of College English Teaching in Heilongjiang Province, China? In T. D. Neimann & U. M. Stelson (Eds.), *Challenges and Opportunities in Global Approaches to Education* (pp. 95–114). IGI Global. doi:10.4018/978-1-5225-9775-9.ch006

Bajit, A. (2020). A Perceptually Optimized Foveation Wavelet Visible Difference Predictor Quality Metric Based on Psychovisual Properties of the Human Visual System (HVS). In M. Sarfraz (Ed.), *Advancements in Computer Vision Applications in Intelligent Systems and Multimedia Technologies* (pp. 268–281). IGI Global. doi:10.4018/978-1-7998-4444-0.ch014

Baker, A. A. (2020). To Whose Benefit? At What Cost? In M. L. Baran & J. E. Jones (Eds.), *Applied Social Science Approaches to Mixed Methods Research* (pp. 251–260). IGI Global. doi:10.4018/978-1-7998-1025-4.ch011

Banerjee, S., & Wahl, M. F. (2020). Corporate Governance and Values. In R. Brunet-Thornton (Ed.), *Examining Cultural Perspectives in a Globalized World* (pp. 175–207). IGI Global. doi:10.4018/978-1-7998-0214-3.ch008

Baran, M. L. (2020). Mixed Methods Research Design. In M. L. Baran & J. E. Jones (Eds.), *Applied Social Science Approaches to Mixed Methods Research* (pp. 26–52). IGI Global. doi:10.4018/978-1-7998-1025-4.ch002

Baran, M. L., & Jones, J. E. (2020). Developing the Research Study. In M. L. Baran & J. E. Jones (Eds.), *Applied Social Science Approaches to Mixed Methods Research* (pp. 262–274). IGI Global. doi:10.4018/978-1-7998-1025-4.ch012

Bardi, M., & Muresan, L. (2020). New Roles for ESP Lecturers. In A. Slapac & S. A. Coppersmith (Eds.), *Beyond Language Learning Instruction* (pp. 181–206). IGI Global. doi:10.4018/978-1-7998-1962-2.ch008

Barnor, J. N., & Patterson, A. A. (2020). Cybercrime Research. In R. Boateng (Ed.), *Handbook of Research on Managing Information Systems in Developing Economies* (pp. 480–502). IGI Global. doi:10.4018/978-1-7998-2610-1.ch024

Barrad, S., Gagnon, S., & Valverde, R. (2020). An Analytics Architecture for Procurement. *International Journal of Information Technologies and Systems Approach, 13*(2), 73–98. doi:10.4018/IJITSA.2020070104

Barrad, S., & Valverde, R. (2020). The Impact of e-Supply Chain Management Systems on Procurement Operations and Cost Reduction in the Electronics Manufacturing Services Industry. *Journal of Media Management and Entrepreneurship, 2*(1), 1–27. doi:10.4018/JMME.2020010101

Barroso, P. M. (2020). Visual Literacy and Visual Rhetoric. In S. Kır (Ed.), *New Media and Visual Communication in Social Networks* (pp. 17–36). IGI Global. doi:10.4018/978-1-7998-1041-4.ch002

Bassiliades, N. (2020). A Tool for Transforming Semantic Web Rule Language to SPARQL Infererecing Notation. *International Journal on Semantic Web and Information Systems, 16*(1), 87–115. doi:10.4018/IJSWIS.2020010105

Bataeva, B., Cheglakova, L., & Melitonyan, O. (2020). Motivation for Responsible Behavior Among Russian SMEs. In A. Kuna-Marszałek & A. Kłysik-Uryszek (Eds.), *CSR and Socially Responsible Investing Strategies in Transitioning and Emerging Economies* (pp. 138–170). IGI Global. doi:10.4018/978-1-7998-2193-9.ch008

Batyashe, N. T. (2021). Deployment of Information Technology Governance Using Architectural Framework. In T. Iyamu (Ed.), *Empowering Businesses With Collaborative Enterprise Architecture Frameworks* (pp. 199–215). IGI Global. doi:10.4018/978-1-5225-8229-8.ch009

Bawa, P. (2020). Game On! *International Journal of Game-Based Learning, 10*(3), 16–46. doi:10.4018/IJGBL.2020070102

Bedford, D. A. (2020). Fair Process in Assessing the Quality of University Faculty. In N. Baporikar & M. Sony (Eds.), *Quality Management Principles and Policies in Higher Education* (pp. 156–187). IGI Global. doi:10.4018/978-1-7998-1017-9.ch009

Ben-Lhachemi, N., & Nfaoui, E. H. (2020). Efficient Weighted Semantic Score Based on the Huffman Coding Algorithm and Knowledge Bases for Word Sequences Embedding. *International Journal on Semantic Web and Information Systems, 16*(2), 126–142. doi:10.4018/IJSWIS.2020040107

Ben Mahmoud, C., Azaiez, I., & Bettahar, F. (2020). Multi-Intentional Description of Learning Semantic Web Services. *International Journal on Semantic Web and Information Systems, 16*(2), 108–125. doi:10.4018/IJSWIS.2020040106

Ben Rehouma, M. (2020). Exploring the Role of Participation in Government Employees' Adoption of IT. *International Journal of Public Administration in the Digital Age, 7*(1), 33–46. doi:10.4018/IJPADA.2020010103

Benfares, C., Akhrif, O., El Bouzekri El Idrissi, Y., & Hamid, K. (2020). Multi-Criteria Decision Making Semantic for Mental Healthcare. *International Journal of Smart Security Technologies, 7*(1), 58–71. doi:10.4018/IJSST.2020010105

Bertrand, M., & Namukasa, I. K. (2020). Integrating Computational Thinking and Mathematics. In J. Keengwe & P. Wachira (Eds.), *Handbook of Research on Integrating Computer Science and Computational Thinking in K-12 Education* (pp. 69–91). IGI Global. doi:10.4018/978-1-7998-1479-5.ch005

Betts, A. L., & Rothschild, M. K. (2020). Massively Multiplayer Online Games as Spaces for Metacognition and Self-Regulated Learning. In M. Farber (Ed.), *Global Perspectives on Gameful and Playful Teaching and Learning* (pp. 78–104). IGI Global. doi:10.4018/978-1-7998-2015-4.ch004

Bhattacharyya, S. S., & Atre, S. (2020). Additive Manufacturing Technology. *International Journal of Asian Business and Information Management, 11*(1), 1–20. doi:10.4018/IJABIM.2020010101

Bloom, L., Noltensmeyer, C. J., Hahn, S. A., Rush, C. B., Heidlebaugh-Buskey, P., & Westbrook, T. M. (2020). Pedagogical Considerations in Teaching Implicit Bias. *International Journal of Bias, Identity and Diversities in Education, 5*(1), 46–63. doi:10.4018/IJBIDE.2020010103

Boachie, C. (2020). Corporate Social Responsibility and Financial Performance in the Banking Industry in Ghana. *International Journal of Sustainable Entrepreneurship and Corporate Social Responsibility, 5*(2), 86–104. doi:10.4018/IJSECSR.2020070106

Bonfanti, A., Canestrino, R., Castellani, P., & Vigolo, V. (2020). The In-Store Shopping Experience. In F. Musso & E. Druica (Eds.), *Handbook of Research on Retailing Techniques for Optimal Consumer Engagement and Experiences* (pp. 110–141). IGI Global. doi:10.4018/978-1-7998-1412-2.ch006

Bouafif, H., Kamoun, F., & Iqbal, F. (2020). Towards a Better Understanding of Drone Forensics. *International Journal of Digital Crime and Forensics, 12*(1), 35–57. doi:10.4018/IJDCF.2020010103

Boubenia, M., Belkhir, A., & Bouyakoub, F. M. (2020). Combining Linked Open Data Similarity and Relatedness for Cross OSN Recommendation. *International Journal on Semantic Web and Information Systems, 16*(2), 59–90. doi:10.4018/IJSWIS.2020040104

Bouras, C., Kokkinos, V., & Michos, E. (2020). A Resource-Efficient Approach on User Association in 5G Networks Using Downlink and Uplink Decoupling. *International Journal of Wireless Networks and Broadband Technologies, 9*(2), 43–59. doi:10.4018/IJWNBT.2020070103

Brandão, A. M., & Machado, S. M. (2020). How Consumers Respond to Editorial Communication Strategies. In S. M. Loureiro & H. R. Kaufmann (Eds.), *Exploring the Power of Electronic Word-of-Mouth in the Services Industry* (pp. 364–389). IGI Global. doi:10.4018/978-1-5225-8575-6.ch020

Brown, C. (2020). *Research Design and Methodology*. IGI Global. doi:10.4018/978-1-7998-1639-3.ch004

Burrell, D. N., Bhargava, N., Springs, D., Dawson, M., Burton, S. L., Anderson, D. P., & Wright, J. B. (2020). Adopting Organizational Cultural Changes Concerning Whistle-Blowing in Healthcare Around Information Security in the "Internet of Things" World. *International Journal of Hyperconnectivity and the Internet of Things*, *4*(1), 13–28. doi:10.4018/IJHIoT.2020010102

Burrell, D. N., Burton, S. L., Springs, D., Wright, J. B., Harmon, M. D., & Lewis, E. J. (2020). A Qualitative Research Assessment of the MBA for 2020 and Beyond. *International Journal of Smart Education and Urban Society*, *11*(3), 23–41. doi:10.4018/IJSEUS.2020070103

Burton, T. R. (2020). Using Social Media to Organize a Marginalized Community. In M. L. Baran & J. E. Jones (Eds.), *Applied Social Science Approaches to Mixed Methods Research* (pp. 216–249). IGI Global. doi:10.4018/978-1-7998-1025-4.ch010

Byrd, M. Y., & Hughes, C. (2021). Re-Conceptualizing Diversity Management. In C. Hughes (Ed.), *Implementation Strategies for Improving Diversity in Organizations* (pp. 39–74). IGI Global. doi:10.4018/978-1-7998-4745-8.ch002

Call-Cummings, M., Hauber-Özer, M., & Rainey, J. (2020). Preparing Critical Educators and Community-Engaged Scholars Through Participatory Action Research. In A. S. Zimmerman (Ed.), *Preparing Students for Community-Engaged Scholarship in Higher Education* (pp. 1–20). IGI Global. doi:10.4018/978-1-7998-2208-0.ch001

Camilleri, M. A. (2020). The Market for Socially Responsible Investments. In A. Kuna-Marszałek & A. Kłysik-Uryszek (Eds.), *CSR and Socially Responsible Investing Strategies in Transitioning and Emerging Economies* (pp. 171–188). IGI Global. doi:10.4018/978-1-7998-2193-9.ch009

Campos, A. C., & Almeida, S. (2020). Operant Resources of Young Independent Travelers. In P. Pinto & M. Guerreiro (Eds.), *Handbook of Research on Resident and Tourist Perspectives on Travel Destinations* (pp. 286–312). IGI Global. doi:10.4018/978-1-7998-3156-3.ch014

Çapraz, B. (2020). It Fits My Business. In A. Kuna-Marszałek & A. Kłysik-Uryszek (Eds.), *CSR and Socially Responsible Investing Strategies in Transitioning and Emerging Economies* (pp. 237–262). IGI Global. doi:10.4018/978-1-7998-2193-9.ch012

Carlson, K., & Valentin, R. (2020). The Gamification of Code. In L. Kyei-Blankson, E. Ntuli, & M. A. Nur-Awaleh (Eds.), *Emerging Techniques and Applications for Blended Learning in K-20 Classrooms* (pp. 88–113). IGI Global. doi:10.4018/978-1-7998-0242-6.ch005

Castañeda, W. A. (2020). Multi-Criteria Decision-Making Process for Decentralized Healthcare Services Assessment. In A. Behl (Ed.), *Multi-Criteria Decision Analysis in Management* (pp. 85–110). IGI Global. doi:10.4018/978-1-7998-2216-5.ch005

Chaker, J., & Khaldi, M. (2020). A New Semantic Description for Multimedia and Learning Objects. In M. A. Tadlaoui & M. Khaldi (Eds.), *Personalization and Collaboration in Adaptive E-Learning* (pp. 195–205). IGI Global. doi:10.4018/978-1-7998-1492-4.ch009

Chaterera-Zambuko, F. (2020). The Use of Grounded Theory in Researching Information Centres. In P. Ngulube (Ed.), *Handbook of Research on Connecting Research Methods for Information Science Research* (pp. 139–158). IGI Global. doi:10.4018/978-1-7998-1471-9.ch008

Chebil, W., Wedyan, M. O., Lu, H., & Elshaweesh, O. G. (2020). Context-Aware Personalized Web Search Using Navigation History. *International Journal on Semantic Web and Information Systems*, *16*(2), 91–107. doi:10.4018/IJSWIS.2020040105

Chergui, M. E., & Benslimane, S. M. (2020). Towards a BPMN Security Extension for the Visualization of Cyber Security Requirements. *International Journal of Technology Diffusion*, *11*(2), 1–17. doi:10.4018/IJTD.2020040101

Cheung, H. W., Bhowmik, M. K., Kennedy, K. J., Ng, H. Y., & Hue, M. T. (2020). Religious and Ethnic Identification of Minoritized Youth in Hong Kong. In E. J. Delgado-Algarra & J. M. Cuenca-López (Eds.), *Handbook of Research on Citizenship and Heritage Education* (pp. 332–351). IGI Global. doi:10.4018/978-1-7998-1978-3.ch016

Choi, Y. (2020). Corporate Psychopathy and Internet Video Game Addiction in the Workplace. *International Journal of Gaming and Computer-Mediated Simulations*, *12*(3), 18–36. doi:10.4018/IJGCMS.2020070102

Choi, Y., & Kim, H. (2020). Exploring the Issues for the Success of Multichannel Network Businesses in Korea. *Journal of Global Information Management*, *28*(2), 90–110. doi:10.4018/JGIM.2020040105

Chukwuere, J. E., & Chukwuere, P. C. (2020). Guide for Postgraduate Student Research and Publications. In P. Ngulube (Ed.), *Handbook of Research on Connecting Research Methods for Information Science Research* (pp. 529–550). IGI Global. doi:10.4018/978-1-7998-1471-9.ch027

Confente, I., Vigolo, V., & Brunetti, F. (2020). The Role of WOM in Affecting the Intention to Purchase Online. In S. M. Loureiro & H. R. Kaufmann (Eds.), *Exploring the Power of Electronic Word-of-Mouth in the Services Industry* (pp. 317–333). IGI Global. doi:10.4018/978-1-5225-8575-6.ch017

Cooke, L., Schugar, J., Schugar, H., Penny, C., & Bruning, H. (2020). Can Everyone Code? In J. S. Mitchell & E. N. Vaughn (Eds.), *Participatory Literacy Practices for P-12 Classrooms in the Digital Age* (pp. 163–183). IGI Global. doi:10.4018/978-1-7998-0000-2.ch009

Cosgrove, P. B. (2020). Implications of Mixing Methods. In M. L. Baran & J. E. Jones (Eds.), *Applied Social Science Approaches to Mixed Methods Research* (pp. 1–24). IGI Global. doi:10.4018/978-1-7998-1025-4.ch001

Costa, J., Vasconcelos, A., & Fragoso, B. (2020). An Enterprise Architecture Approach for Assessing the Alignment Between Projects and Goals. *International Journal of Information Technology Project Management, 11*(3), 55–76. doi:10.4018/IJITPM.2020070104

Crain-Dorough, M. (2020). Considering the Source. In M. L. Baran & J. E. Jones (Eds.), *Applied Social Science Approaches to Mixed Methods Research* (pp. 54–85). IGI Global. doi:10.4018/978-1-7998-1025-4.ch003

Cullen, T. A. (2020). Connecting With Industry in Computer Science Education. In J. Keengwe & P. Wachira (Eds.), *Handbook of Research on Integrating Computer Science and Computational Thinking in K-12 Education* (pp. 92–106). IGI Global. doi:10.4018/978-1-7998-1479-5.ch006

Dantas, J. R., & Farias, P. P. (2020). An Architecture for Restful Web Service Discovery Using Semantic Interfaces. *International Journal on Semantic Web and Information Systems, 16*(1), 1–24. doi:10.4018/IJSWIS.2020010101

Dasti, R., Farhan, R., Naqvi, T. F., Sitwat, A., & Bokharey, I. Z. (2020). Ethical and Moral Dilemmas in Psychotherapy. In A. Bagasra & M. Mackinem (Eds.), *Working With Muslim Clients in the Helping Professions* (pp. 242–258). IGI Global. doi:10.4018/978-1-7998-0018-7.ch013

Deka, P. (2021). *Genomics, Proteomics, and Metabolomics*. IGI Global. doi:10.4018/978-1-7998-4312-2.ch011

Deka, P. (2021). *Molecular Markers for Phylogenetic Studies and Germplasm Conservation*. IGI Global. doi:10.4018/978-1-7998-4312-2.ch005

Deka, P. (2021). *Plant DNA Barcoding*. IGI Global. doi:10.4018/978-1-7998-4312-2.ch006

Delgoshaei, A., & Ali, A. (2020). A Hybrid Ant Colony Optimization and Simulated Annealing Algorithm for Multi-Objective Scheduling of Cellular Manufacturing Systems. *International Journal of Applied Metaheuristic Computing*, *11*(3), 1–40. doi:10.4018/IJAMC.2020070101

Devlin, M. (2020). Nine Elements of the Scholarship of Discovery. In L. Ling & P. Ling (Eds.), *Emerging Methods and Paradigms in Scholarship and Education Research* (pp. 84–95). IGI Global. doi:10.4018/978-1-7998-1001-8.ch005

Dhamija, P., Bedi, M., & Gupta, M. (2020). Industry 4.0 and Supply Chain Management. *International Journal of Business Analytics*, *7*(1), 1–23. doi:10.4018/IJBAN.2020010101

Dhir, S., & Dhir, S. (2020). Factor Affecting Innovation Performance of Manufacturing Firms. *International Journal of Asian Business and Information Management*, *11*(3), 85–100. doi:10.4018/IJABIM.2020070106

Dongardive, P. B. (2020). Challenges and Opportunities in Building a Successful Digital Library in Developing Countries. *International Journal of ICT Research in Africa and the Middle East*, *9*(1), 24–49. doi:10.4018/IJICTRAME.2020010102

Doroshenko, A., & Yatsenko, O. (2021). *Software Design Based on Using Ontologies and Algorithm Algebra*. IGI Global. doi:10.4018/978-1-5225-9384-3.ch007

Doughty, H. A. (2020). Administrative Ethics in the Corporate College. In V. Wang (Ed.), *Handbook of Research on Ethical Challenges in Higher Education Leadership and Administration* (pp. 131–155). IGI Global. doi:10.4018/978-1-7998-4141-8.ch008

Du, J. (2020). Constructing the Internet Panoptic-Fortification. In V. Kumar & G. Malhotra (Eds.), *Examining the Roles of IT and Social Media in Democratic Development and Social Change* (pp. 211–249). IGI Global. doi:10.4018/978-1-7998-1791-8.ch009

Dumbu, E. (2020). Faculty Perspective Regarding Challenges in Implementing Quality Assurance in Zimbabwean Higher Education Institutions. In M. Sony, K. T. Karingada, & N. Baporikar (Eds.), *Quality Management Implementation in Higher Education* (pp. 372–385). IGI Global. doi:10.4018/978-1-5225-9829-9.ch018

Ensslin, L., Mussi, C. C., Dutra, A., Ensslin, S. R., & Demetrio, S. N. (2020). Management Support Model for Information Technology Outsourcing. *Journal of Global Information Management, 28*(3), 123–147. doi:10.4018/JGIM.2020070107

Eroglu, S., & Kucun, N. T. (2020). Traditional Market Research and Neuromarketing Research. In D. Atli (Ed.), *Analyzing the Strategic Role of Neuromarketing and Consumer Neuroscience* (pp. 146–167). IGI Global. doi:10.4018/978-1-7998-3126-6.ch008

Etike, S. (2020). Thinking Outside the Boxes. In S. Karlidag & S. Bulut (Eds.), *Handbook of Research on the Political Economy of Communications and Media* (pp. 227–241). IGI Global. doi:10.4018/978-1-7998-3270-6.ch012

Ettaib, K., & Khaldi, M. (2020). Contribution of Ontologies and Utilities of Interoperability Standardization in E-Learning. In M. A. Tadlaoui & M. Khaldi (Eds.), *Personalization and Collaboration in Adaptive E-Learning* (pp. 176–194). IGI Global. doi:10.4018/978-1-7998-1492-4.ch008

Ettazi, W., Nassar, M., & Hafiddi, H. (2020). CATS-CAE Reflective Middleware Framework for Adapting Context-Aware Transactional Services. *International Journal of Web Services Research, 17*(2), 40–58. doi:10.4018/IJWSR.2020040103

Eyupoglu, A., & Kaya, T. (2020). E-Government Awareness and Adoption by the Residents. *International Journal of Public Administration in the Digital Age, 7*(2), 1–22. doi:10.4018/IJPADA.2020040101

Fallah, S., Reynolds, B., & Murawski, W. (2020). *Researching a Unique CLD Population*. IGI Global. doi:10.4018/978-1-7998-2069-7.ch006

Farid, S., & Mostari, M. (2020). Inclusive Approach to Education for Children With Disabilities. *International Journal of Teacher Education and Professional Development, 3*(1), 121–139. doi:10.4018/IJTEPD.2020010108

Fatima, T. (2020). Impact of Employees' Perceived Corporate Social Responsibility on Organizational Citizenship Behavior. *International Journal of Customer Relationship Marketing and Management, 11*(3), 25–38. doi:10.4018/IJCRMM.2020070102

Fayoumi, A. G., & Hajjar, A. F. (2020). Advanced Learning Analytics in Academic Education. *International Journal on Semantic Web and Information Systems, 16*(3), 70–87. doi:10.4018/IJSWIS.2020070105

Fedoruk, L. M., & Mikita, K. (2020). The Role of Educational Developer in Supporting Research Ethics in SoTL. In R. C. Plews & M. L. Amos (Eds.), *Evidence-Based Faculty Development Through the Scholarship of Teaching and Learning (SoTL)* (pp. 221–236). IGI Global. doi:10.4018/978-1-7998-2212-7.ch012

Ferrari, F. (2020). Investigating Inner Dynamics in Family Firms. In C. Camisón & T. González (Eds.), *Competitiveness, Organizational Management, and Governance in Family Firms* (pp. 330–355). IGI Global. doi:10.4018/978-1-7998-1655-3.ch014

Ficko, M., Berus, L., Palčič, I., & Klančnik, S. (2020). Design of Facility Layout for Industry 4.0. In I. Karabegović, A. Kovačević, L. Banjanović-Mehmedović, & P. Dašić (Eds.), *Handbook of Research on Integrating Industry 4.0 in Business and Manufacturing* (pp. 101–126). IGI Global. doi:10.4018/978-1-7998-2725-2.ch005

Flore, A., & Marx Goméz, J. (2021). Requirements on Dimensions for a Maturity Model for Smart Grids Based on Two Case Studies. In M. Mora, J. M. Gómez, R. V. O'Connor, & A. Buchalcevová (Eds.), *Balancing Agile and Disciplined Engineering and Management Approaches for IT Services and Software Products* (pp. 271–292). IGI Global. doi:10.4018/978-1-7998-4165-4.ch014

Flotyński, J., Malamos, A. G., Brutzman, D., Hamza-Lup, F. G., Polys, N. F., Sikos, L. F., & Walczak, K. (2020). Recent Advances in Web3D Semantic Modeling. In A. Voulodimos & A. Doulamis (Eds.), *Recent Advances in 3D Imaging, Modeling, and Reconstruction* (pp. 23–49). IGI Global. doi:10.4018/978-1-5225-5294-9.ch002

França, R. P., Iano, Y., Monteiro, A. C., & Arthur, R. (2020). A Review on the Technological and Literary Background of Multimedia Compression. In B. B. Gupta & D. Gupta (Eds.), *Handbook of Research on Multimedia Cyber Security* (pp. 1–20). IGI Global. doi:10.4018/978-1-7998-2701-6.ch001

Fredrickson, J. (2020). Billing and Review Perspectives in Healthcare. In J. Khuntia, X. Ning, & M. Tanniru (Eds.), *Theory and Practice of Business Intelligence in Healthcare* (pp. 18–67). IGI Global. doi:10.4018/978-1-7998-2310-0.ch002

Freeman, C. (2020). Leader Launch. In J. Stefaniak (Ed.), *Cases on Instructional Design and Performance Outcomes in Medical Education* (pp. 235–260). IGI Global. doi:10.4018/978-1-7998-5092-2.ch012

Froehlich, T. J. (2020). Ten Lessons for the Age of Disinformation. In K. Dalkir & R. Katz (Eds.), *Navigating Fake News, Alternative Facts, and Misinformation in a Post-Truth World* (pp. 36–88). IGI Global. doi:10.4018/978-1-7998-2543-2.ch003

Fukuzawa, S., Ashbourne, D., & Rawle, F. (2020). Overcoming Challenges to Impactful SoTL. In R. C. Plews & M. L. Amos (Eds.), *Evidence-Based Faculty Development Through the Scholarship of Teaching and Learning (SoTL)* (pp. 366–387). IGI Global. doi:10.4018/978-1-7998-2212-7.ch019

Fülöp, M. T., Cordoş, G. S., & Măgdaş, N. (2020). Importance of Ethics and Education to Understand the Audit Mission. In I. Oncioiu, G. Cokins, S. Căpuşneanu, & D. I. Topor (Eds.), *Management Accounting Standards for Sustainable Business Practices* (pp. 209–226). IGI Global. doi:10.4018/978-1-7998-0178-8.ch011

Gafour, Y., & Berrabah, D. (2020). New Approach to Improve the Classification Process of Multi-Class Objects. *International Journal of Organizational and Collective Intelligence*, *10*(2), 1–19. doi:10.4018/IJOCI.2020040101

Galli, B. J. (2020). Continuous Improvement, Six Sigma and Risk Management. *International Journal of Strategic Engineering*, *3*(2), 1–23. doi:10.4018/IJoSE.2020070101

Galli, B. J. (2020). Continuous Technological Improvement Using Systems Engineering Principles to Achieve Sustainability. *International Journal of System Dynamics Applications*, *9*(3), 1–25. doi:10.4018/IJSDA.2020070101

Galli, B. J. (2020). Economic Decision Making and Risk Management in a Project Management Environment. *International Journal of Applied Logistics*, *10*(2), 1–24. doi:10.4018/IJAL.2020070101

Galli, B. J. (2020). Effective and Ineffective Statistical Analysis Tools in Project Management Environments. *International Journal of Applied Logistics*, *10*(1), 41–57. doi:10.4018/IJAL.2020010104

Galli, B. J. (2020). Measurement System Analysis and System Thinking in Six Sigma. *International Journal of System Dynamics Applications*, *9*(1), 44–62. doi:10.4018/IJSDA.2020010103

Garg, P. (2020). Impact of Digitization on Commercial Banking Services. In G. Jain, H. Singh, S. Akter, A. Munjal, & H. S. Grewal (Eds.), *Technological Innovations for Sustainability and Business Growth* (pp. 71–90). IGI Global. doi:10.4018/978-1-5225-9940-1.ch006

Garip, S. B., Saglar Onay, N., & Garip, E. (2020). Re-Coding Homes as a Flexible Design Approach for Living Environments. In L. Crespi (Ed.), *Cultural, Theoretical, and Innovative Approaches to Contemporary Interior Design* (pp. 284–306). IGI Global. doi:10.4018/978-1-7998-2823-5.ch013

Garozzo, R., Pino, C., Santagati, C., & Spampinato, C. (2020). Harnessing the Power of Artificial Intelligence for Modelling and Understanding Cultural Heritage Data. In C. M. Bolognesi & C. Santagati (Eds.), *Impact of Industry 4.0 on Architecture and Cultural Heritage* (pp. 357–376). IGI Global. doi:10.4018/978-1-7998-1234-0.ch015

Gayathri, K., Easwarakumar, K., & Elias, S. (2020). Fuzzy Ontology Based Activity Recognition for Assistive Health Care Using Smart Home. *International Journal of Intelligent Information Technologies*, *16*(1), 17–31. doi:10.4018/IJIIT.2020010102

Genel, Z. (2020). The New Horizon of Public Relations. In R. C. Ho (Ed.), *Strategies and Tools for Managing Connected Consumers* (pp. 208–222). IGI Global. doi:10.4018/978-1-5225-9697-4.ch012

Ghorbel, F., Hamdi, F., & Métais, E. (2020). Dealing with Precise and Imprecise Temporal Data in Crisp Ontology. *International Journal of Information Technology and Web Engineering*, *15*(2), 30–49. doi:10.4018/IJITWE.2020040102

Giannouli, V. (2020). Business Ethics in Healthcare. In F. S. Topor (Ed.), *Examining Ethics and Intercultural Interactions in International Relations* (pp. 138–167). IGI Global. doi:10.4018/978-1-7998-2377-3.ch006

Gillenson, M. L., Stafford, T. F., Zhang, X., & Shi, Y. (2020). Use of Qualitative Research to Generate a Function for Finding the Unit Cost of Software Test Cases. *Journal of Database Management, 31*(2), 42–63. doi:10.4018/JDM.2020040103

Giridharan, B. (2020). Education Research and Scholarship in an Intercultural Context. In L. Ling & P. Ling (Eds.), *Emerging Methods and Paradigms in Scholarship and Education Research* (pp. 118–133). IGI Global. doi:10.4018/978-1-7998-1001-8.ch007

Glezou, K. V. (2020). Fostering Computational Thinking and Creativity in Early Childhood Education. In S. Papadakis & M. Kalogiannakis (Eds.), *Mobile Learning Applications in Early Childhood Education* (pp. 324–347). IGI Global. doi:10.4018/978-1-7998-1486-3.ch016

Gökçe, A. T., & Durante, R. (2020). Ethical Challenges. In V. Wang (Ed.), *Handbook of Research on Ethical Challenges in Higher Education Leadership and Administration* (pp. 172–202). IGI Global. doi:10.4018/978-1-7998-4141-8.ch010

Gomes, S. B., Santoro, F. M., & Mira da Silva, M. (2020). An Ontology for BPM in Digital Transformation and Innovation. *International Journal of Information System Modeling and Design, 11*(2), 52–77. doi:10.4018/IJISMD.2020040103

Green, M., & McNair, C. L. (2020). Steamsational Writing. In P. M. Sullivan, J. L. Lantz, & B. A. Sullivan (Eds.), *Handbook of Research on Integrating Digital Technology With Literacy Pedagogies* (pp. 175–191). IGI Global. doi:10.4018/978-1-7998-0246-4.ch008

Greene, M., Clarke-Hagan, D., & Curran, M. (2020). Achieving Smarter Buildings and More Efficient Facilities Management. *International Journal of Digital Innovation in the Built Environment, 9*(2), 1–16. doi:10.4018/IJDIBE.2020070101

Gregorka, L., Silva, S., & Silva, C. (2020). Employer Branding Practices Amongst the Most Attractive Employers of IT and Engineering Sector. *International Journal of Human Capital and Information Technology Professionals, 11*(1), 1–16. doi:10.4018/IJHCITP.2020010101

Guerreiro, J. (2020). Do We Really Care About Artificial Intelligence? In S. M. Loureiro (Ed.), *Managerial Challenges and Social Impacts of Virtual and Augmented Reality* (pp. 234–251). IGI Global. doi:10.4018/978-1-7998-2874-7.ch014

Guerreiro, S. L. (2020). On Ontology of Integration Between Access Control and Business Process Deep Structure. In N. Prakash & D. Prakash (Eds.), *Novel Approaches to Information Systems Design* (pp. 226–246). IGI Global. doi:10.4018/978-1-7998-2975-1.ch010

Gulzar, Z., Raj, L. A., & Leema, A. A. (2020). Towards Increasing the Efficiency of E-Learning Systems Using Recommendation System Approach. In C. Bhatt, P. S. Sajja, & S. Liyanage (Eds.), *Utilizing Educational Data Mining Techniques for Improved Learning* (pp. 35–55). IGI Global. doi:10.4018/978-1-7998-0010-1.ch003

Gupta, V., Chauhan, D. S., & Hanne, T. (2020). Flexible Higher Education Through Swayam. In L. Kyei-Blankson, E. Ntuli, & J. Blankson (Eds.), *Handbook of Research on Creating Meaningful Experiences in Online Courses* (pp. 59–65). IGI Global. doi:10.4018/978-1-7998-0115-3.ch005

Gupta, Y., & Shome, S. (2020). Social Media Advertisements and Buying Behaviour. *International Journal of Online Marketing, 10*(3), 48–61. doi:10.4018/IJOM.2020070104

Hadjistassou, S., Avgousti, M. I., & Allen, C. (2020). Pre-Service Teachers' and Instructors' Reflections on Virtually Afforded Feedback During a Distant Teaching Practicum Experience. *International Journal of Teacher Education and Professional Development, 3*(1), 42–59. doi:10.4018/IJTEPD.2020010103

Hai-Jew, S. (2020). Designing Integrated Learning Paths for Individual Lifelong Learners and/or Small Groups. In S. Hai-Jew (Ed.), *Form, Function, and Style in Instructional Design* (pp. 68–104). IGI Global. doi:10.4018/978-1-5225-9833-6.ch005

Hai-Jew, S. (2020). Exploring Social Imagery Depictions of the Meeting of Human Needs in the Global South. In F. P. Endong (Ed.), *Deconstructing Images of the Global South Through Media Representations and Communication* (pp. 87–109). IGI Global. doi:10.4018/978-1-5225-9821-3.ch004

Hai-Jew, S. (2020). Global North-Global South Relations Around a Notional U.S.-Mexico "Border Wall". In F. P. Endong (Ed.), *Deconstructing Images of the Global South Through Media Representations and Communication* (pp. 296–344). IGI Global. doi:10.4018/978-1-5225-9821-3.ch013

Hai-Jew, S. (2020). Image on the Street Is. In F. P. Endong (Ed.), *Deconstructing Images of the Global South Through Media Representations and Communication* (pp. 1–45). IGI Global. doi:10.4018/978-1-5225-9821-3.ch001

Hai-Jew, S. (2020). Person-to-Person Direct Transactions Between the Global North <-> Global South in Social Videos. In F. P. Endong (Ed.), *Deconstructing Images of the Global South Through Media Representations and Communication* (pp. 345–382). IGI Global. doi:10.4018/978-1-5225-9821-3.ch014

Hai-Jew, S. (2020). Shapes of Threats in the Global South. In F. P. Endong (Ed.), *Deconstructing Images of the Global South Through Media Representations and Communication* (pp. 110–165). IGI Global. doi:10.4018/978-1-5225-9821-3.ch005

Halaweh, M., Elbahi, M., Kamel, A., Kabha, R., & Yousef, R. (2020). Exploring the Use of Social Media Platforms by Public Universities. *International Journal of Web Portals, 12*(2), 41–56. doi:10.4018/IJWP.2020070103

Hayes, C., & Corrie, I. (2020). Learner-Centred Pedagogy Framing Authentic Identity and Positionality in Higher Education. In Y. Inoue-Smith & T. McVey (Eds.), *Optimizing Higher Education Learning Through Activities and Assessments* (pp. 77–95). IGI Global. doi:10.4018/978-1-7998-4036-7.ch005

Heasley, T. (2020). Addressing the Need for Anger Management in Young People. In S. G. Taukeni (Ed.), *Addressing Multicultural Needs in School Guidance and Counseling* (pp. 33–63). IGI Global. doi:10.4018/978-1-7998-0319-5.ch003

Henschke, J. A. (2020). Leadership Ethics in Higher Education Administration. In V. Wang (Ed.), *Handbook of Research on Ethical Challenges in Higher Education Leadership and Administration* (pp. 332–360). IGI Global. doi:10.4018/978-1-7998-4141-8.ch018

Hernández-Sampieri, R., & Méndez-Valencia, S. (2020). Research Methods for the Study of Small and Medium-Sized Enterprises. In R. Perez-Uribe, D. Ocampo-Guzman, C. Salcedo-Perez, L. Piñeiro-Cortes, & M. D. Ramirez-Salazar (Eds.), *Handbook of Research on Increasing the Competitiveness of SMEs* (pp. 125–151). IGI Global. doi:10.4018/978-1-5225-9425-3.ch006

Hilal, S. (2020). A Survey of Ontology-Based Frameworks for Sustainable Supply Chain Interoperability and Collaboration. In A. Awasthi & K. Grzybowska (Eds.), *Handbook of Research on Interdisciplinary Approaches to Decision Making for Sustainable Supply Chains* (pp. 459–475). IGI Global. doi:10.4018/978-1-5225-9570-0.ch021

Hilles, S. M., & Maidaniuk, V. P. (2020). SOFM for Image Compression Based on Spatial Frequency Band-Pass Filter and Vector Quantization. In S. M. Bilan & S. I. Al-Zoubi (Eds.), *Handbook of Research on Intelligent Data Processing and Information Security Systems* (pp. 388–403). IGI Global. doi:10.4018/978-1-7998-1290-6.ch016

Hozdić, E. (2020). An Ontology-Based Approach for Production Planning and Control of Cyber-Physical Production Systems. In I. Karabegović, A. Kovačević, L. Banjanović-Mehmedović, & P. Dašić (Eds.), *Handbook of Research on Integrating Industry 4.0 in Business and Manufacturing* (pp. 274–300). IGI Global. doi:10.4018/978-1-7998-2725-2.ch013

Huang, C. C., Lee, C. E., & Thing, V. L. (2020). A Novel Video Forgery Detection Model Based on Triangular Polarity Feature Classification. *International Journal of Digital Crime and Forensics, 12*(1), 14–34. doi:10.4018/IJDCF.2020010102

Huda, S. S., Akhtar, A., Kabir, M. H., & Faruq, O. (2021). Advertisement Practices in Bangladesh. In E. Esiyok (Ed.), *Handbook of Research on New Media Applications in Public Relations and Advertising* (pp. 63–100). IGI Global. doi:10.4018/978-1-7998-3201-0.ch005

Huett, K. C., D'Alba, A., & Neubauer, B. (2020). The Design, Implementation, and Evaluation of a Graduate Course to Prepare Teachers and School Librarians to Lead K-12 Computational Thinking. In J. Keengwe & P. Wachira (Eds.), *Handbook of Research on Integrating Computer Science and Computational Thinking in K-12 Education* (pp. 243–268). IGI Global. doi:10.4018/978-1-7998-1479-5.ch013

Hussain, I., & Asif, M. (2020). Detection of Anomalous Transactions in Mobile Payment Systems. *International Journal of Data Analytics, 1*(2), 58–66. doi:10.4018/IJDA.2020070105

Hutchinson, A., & Stoeckle, A. (2020). Using Mid-Semester Assessment Programs (MAPs) as a Catalyst for the Scholarship of Teaching and Learning (SoTL). In R. C. Plews & M. L. Amos (Eds.), *Evidence-Based Faculty Development Through the Scholarship of Teaching and Learning (SoTL)* (pp. 181–200). IGI Global. doi:10.4018/978-1-7998-2212-7.ch010

Ibarra, L., Soriano, A., Ponce, P., & Molina, A. (2020). The Wit-Learning Methodology as a Means for Research Skills Acquisition. In M. K. Habib (Ed.), *Revolutionizing Education in the Age of AI and Machine Learning* (pp. 196–222). IGI Global. doi:10.4018/978-1-5225-7793-5.ch010

Ibrahim, A. M., Pate, U. A., & Usman, A. M. (2020). Silencing the Media and Chaining the Watchdog. In S. Jamil (Ed.), *Handbook of Research on Combating Threats to Media Freedom and Journalist Safety* (pp. 214–240). IGI Global. doi:10.4018/978-1-7998-1298-2.ch012

Ibrahim, F., Ali, D. N., & Besar, N. S. (2020). Accounting Information Systems (AIS) in SMEs. *International Journal of Asian Business and Information Management, 11*(2), 51–67. doi:10.4018/IJABIM.2020040104

Isika, N., Mendoza, A., & Bosua, R. (2020). An Affordance Perspective on the Enabling and Disruptive Effects of Social Media Tools on Self-Management of Chronic Illness. In K. Sandhu (Ed.), *Opportunities and Challenges in Digital Healthcare Innovation* (pp. 36–56). IGI Global. doi:10.4018/978-1-7998-3274-4.ch003

Islam, M. N., Ahmed, M. A., & Islam, A. N. (2020). Chakuri-Bazaar: A Mobile Application for Illiterate and Semi-Literate People for Searching Employment. *International Journal of Mobile Human Computer Interaction, 12*(2), 22–39. doi:10.4018/IJMHCI.2020040102

Ivanov, B., Trif, V., & Trif, A. (2020). Assessment and Paradigms. In V. Trif (Ed.), *Analyzing Paradigms Used in Education and Educational Psychology* (pp. 121–143). IGI Global. doi:10.4018/978-1-7998-1427-6.ch006

Iyamu, T. (2020). A Case for Applying Activity Theory in IS Research. *Information Resources Management Journal, 33*(1), 1–15. doi:10.4018/IRMJ.2020010101

Iyioke, I. C. (2020). Cognitive vs. Social Constructivist Learning for Research and Training on the Angoff Method. In Ş. Orakcı (Ed.), *Paradigm Shifts in 21st Century Teaching and Learning* (pp. 181–201). IGI Global. doi:10.4018/978-1-7998-3146-4.ch012

Jabłoński, A. (2020). Robustness in the Business Models of the Organizations Embedded in the Circular Economy. In B. Nogalski, A. Szpitter, A. Jabłoński, & M. Jabłoński (Eds.), *Networked Business Models in the Circular Economy* (pp. 19–53). IGI Global. doi:10.4018/978-1-5225-7850-5.ch002

Jackson, M. A. (2020). Regulating AI. In M. Jackson & M. Shelly (Eds.), *Legal Regulations, Implications, and Issues Surrounding Digital Data* (pp. 159–181). IGI Global. doi:10.4018/978-1-7998-3130-3.ch009

Jacobs, L. (2020). Embedding Ethical Principles in the Information Science Research Process. In P. Ngulube (Ed.), *Handbook of Research on Connecting Research Methods for Information Science Research* (pp. 52–79). IGI Global. doi:10.4018/978-1-7998-1471-9.ch004

Jagani, K., Oza, F. V., & Chauhan, H. (2020). Customer Segmentation and Factors Affecting Willingness to Order Private Label Brands. In Y. Arslan (Ed.), *Improving Marketing Strategies for Private Label Products* (pp. 227–253). IGI Global. doi:10.4018/978-1-7998-0257-0.ch011

Jain, R., & Singh, A. K. (2020). Technological Approaches for E-Content Development and Deployment. *International Journal of Information and Communication Technology Education*, *16*(3), 92–112. doi:10.4018/IJICTE.2020070107

Jain, S., Sharma, S., Natterbrede, J. M., & Hamada, M. (2020). Rule-Based Actionable Intelligence for Disaster Situation Management. *International Journal of Knowledge and Systems Science*, *11*(3), 17–32. doi:10.4018/IJKSS.2020070102

Janićijević, N. (2020). The Role of the Accreditation Process in Introducing Innovations Into Higher Education in Serbia. In V. Babić & Z. Nedelko (Eds.), *Handbook of Research on Enhancing Innovation in Higher Education Institutions* (pp. 75–96). IGI Global. doi:10.4018/978-1-7998-2708-5.ch004

Jayapalan, A. (2020). USRP-Based Secure Data Transmission. In B. B. Gupta & D. Gupta (Eds.), *Handbook of Research on Multimedia Cyber Security* (pp. 227–242). IGI Global. doi:10.4018/978-1-7998-2701-6.ch012

Jiang, W., Yang, M., Xie, Y., & Li, Z. (2020). Machine Learning-Based Coding Decision Making in H.265/HEVC CTU Division and Intra Prediction. *International Journal of Mobile Computing and Multimedia Communications*, *11*(2), 41–60. doi:10.4018/IJMCMC.2020040103

Jnr, B. A. (2020). A Holistic Study on Green IT/IS Practices in ICT Departments of Collaborative Enterprise. *International Journal of Social Ecology and Sustainable Development*, *11*(2), 1–26. doi:10.4018/IJSESD.2020040101

Johannessen, M. R. (2020). A Research Design for the Examination of Political Empowerment Through Social Media. In A. M. Solo (Ed.), *Handbook of Research on Politics in the Computer Age* (pp. 45–63). IGI Global. doi:10.4018/978-1-7998-0377-5.ch003

Joseph, T. (2020). Ethics in Organization and Management. *International Journal of Business Strategy and Automation*, *1*(3), 67–74. doi:10.4018/IJBSA.20200701.oa1

Kaale, K. B., & Mgeta, M. B. (2020). Photojournalism Ethics. In O. Oyero (Ed.), *Media and Its Role in Protecting the Rights of Children in Africa* (pp. 149–168). IGI Global. doi:10.4018/978-1-7998-0329-4.ch008

Kaban, A. L., & Boy Ergul, I. (2020). Teachers' Attitudes Towards the Use of Tablets in Six EFL Classrooms. In E. Podovšovnik (Ed.), *Examining the Roles of Teachers and Students in Mastering New Technologies* (pp. 284–298). IGI Global. doi:10.4018/978-1-7998-2104-5.ch015

Kabeyi, M. J. (2020). Project and Program Evaluation Consultancy With Terms of Reference, Challenges, Opportunities, and Recommendations. *International Journal of Project Management and Productivity Assessment*, *8*(2), 47–68. doi:10.4018/IJPMPA.2020070103

Kalkavan, H. (2020). The Importance of Ethics and Corporate Social Responsibility in Financial Markets. In H. Dinçer & S. Yüksel (Eds.), *Handbook of Research on Decision-Making Techniques in Financial Marketing* (pp. 172–186). IGI Global. doi:10.4018/978-1-7998-2559-3.ch008

Kalogiannakis, M., & Papadakis, S. (2020). The Use of Developmentally Mobile Applications for Preparing Pre-Service Teachers to Promote STEM Activities in Preschool Classrooms. In S. Papadakis & M. Kalogiannakis (Eds.), *Mobile Learning Applications in Early Childhood Education* (pp. 82–100). IGI Global. doi:10.4018/978-1-7998-1486-3.ch005

Kamagasako, M., Shishido, N., & Ikuta, S. (2020). Learning Support System for Programming Language Python. In S. Ikuta (Ed.), *Handbook of Research on Software for Gifted and Talented School Activities in K-12 Classrooms* (pp. 149–168). IGI Global. doi:10.4018/978-1-7998-1400-9.ch007

Kammoun, S., Loukil, S., & Loukil, Y. B. (2020). The Impact of Firm Performance and Corporate Governance on Corporate Social Responsibility. In I. S. Paiva & L. C. Carvalho (Eds.), *Conceptual and Theoretical Approaches to Corporate Social Responsibility, Entrepreneurial Orientation, and Financial Performance* (pp. 266–288). IGI Global. doi:10.4018/978-1-7998-2128-1.ch014

Kang, Y., & Yang, K. C. (2020). Employing Digital Reality Technologies in Art Exhibitions and Museums. In G. Guazzaroni & A. S. Pillai (Eds.), *Virtual and Augmented Reality in Education, Art, and Museums* (pp. 139–161). IGI Global. doi:10.4018/978-1-7998-1796-3.ch008

Karademir Coşkun, T., & Alper, A. (2020). An Ecological Approach to Technology Appropriation. In E. Podovšovnik (Ed.), *Examining the Roles of Teachers and Students in Mastering New Technologies* (pp. 79–106). IGI Global. doi:10.4018/978-1-7998-2104-5.ch004

Kari, T., Makkonen, M., Frank, L., & Kettunen, E. (2020). Expectations and Experiences of Implementing a Mobile Secure Communication Application. *International Journal of E-Services and Mobile Applications*, *12*(1), 80–96. doi:10.4018/IJESMA.2020010105

Karolak, H. (2020). Communication Ethics Literacy and Counterterrorism. In F. S. Topor (Ed.), *Examining Ethics and Intercultural Interactions in International Relations* (pp. 224–249). IGI Global. doi:10.4018/978-1-7998-2377-3.ch009

Kashevnik, A., & Teslya, N. (2020). Ontology-Based Coalition Creation by Autonomous Agents in Smart Space. In S. Balandin & E. Balandina (Eds.), *Tools and Technologies for the Development of Cyber-Physical Systems* (pp. 28–50). IGI Global. doi:10.4018/978-1-7998-1974-5.ch002

Kasten, J. E. (2020). Trust, Organizational Decision-Making, and Data Analytics. *International Journal of Business Intelligence Research*, *11*(1), 22–37. doi:10.4018/IJBIR.2020010102

Kellinger, J. J. (2020). Coding Across the Curriculum. In J. Keengwe & G. Onchwari (Eds.), *Handbook of Research on Literacy and Digital Technology Integration in Teacher Education* (pp. 214–227). IGI Global. doi:10.4018/978-1-7998-1461-0.ch012

Kelly, A. (2020). Staff Accessibility and Online Engagement With First-Year Students. *International Journal of Online Pedagogy and Course Design, 10*(1), 48–60. doi:10.4018/IJOPCD.2020010104

Kelly, M. (2020). Teacher's Ontological and Epistemological Beliefs. In L. Ling & P. Ling (Eds.), *Emerging Methods and Paradigms in Scholarship and Education Research* (pp. 135–163). IGI Global. doi:10.4018/978-1-7998-1001-8.ch008

Kettunen, E., Kemppainen, T., Lievonen, M., Makkonen, M., Frank, L., & Kari, T. (2020). Identifying the Ideal Types of Online Shoppers. *International Journal of E-Services and Mobile Applications, 12*(2), 59–78. doi:10.4018/IJESMA.2020040104

Khalid, R., Mir, S. R., Sohail, K., & Tawfik, S. (2020). Cultural Sensitivity. *International Journal of Project Management and Productivity Assessment, 8*(2), 23–46. doi:10.4018/IJPMPA.2020070102

Khallouki, H., Benlamri, R., & Yassine, A. (2020). Comprehensive Ontological Model for Senior Wellness Activity Recognition in Smart Homes. In S. Krit, V. E. Bălaş, M. Elhoseny, R. Benlamri, & M. M. Bălaş (Eds.), *Sensor Network Methodologies for Smart Applications* (pp. 148–167). IGI Global. doi:10.4018/978-1-7998-4381-8.ch007

Khazanchi, P., & Khazanchi, R. (2020). Integration of Educational Software in Teaching Gifted Students in K-12 Classrooms. In S. Ikuta (Ed.), *Handbook of Research on Software for Gifted and Talented School Activities in K-12 Classrooms* (pp. 43–64). IGI Global. doi:10.4018/978-1-7998-1400-9.ch003

Kieu, T. A., & Le, T. D. (2020). Mindfulness Approach to Ethical Consumption. In B. R. Hernández-Sánchez, J. C. Sánchez-García, & A. C. Moreira (Eds.), *Building an Entrepreneurial and Sustainable Society* (pp. 289–304). IGI Global. doi:10.4018/978-1-7998-2704-7.ch015

Korstanje, M. (2020). *The Theory of Risk Perception Geared to Tourism and Hospitality Fields*. IGI Global. doi:10.4018/978-1-7998-0070-5.ch001

Kotsios, P., & Folinas, D. (2020). Analysis and Comparison of Road Freight Transport Cost in 20 European Countries. *International Journal of Applied Logistics, 10*(1), 13–26. doi:10.4018/IJAL.2020010102

Kourtiche, A., Benslimane, S. M., & Hacene, S. B. (2020). OUPIP. *International Journal of Knowledge-Based Organizations, 10*(2), 12–34. doi:10.4018/IJKBO.2020040102

Kovaleva, A. S., Lula, P., & Tuchowski, J. (2020). Ontology-Based Measurement of Study Program Innovativeness in the Area of Economics and Management. In V. Babić & Z. Nedelko (Eds.), *Handbook of Research on Enhancing Innovation in Higher Education Institutions* (pp. 381–407). IGI Global. doi:10.4018/978-1-7998-2708-5.ch017

Kugara, S. L., Mdhluli, T. D., Daswa, T. J., Matshidze, P. E., & Ramavhunga, N. E. (2021). The Significance of Multicultural Methodologies on African Indigenous Knowledge Research. In N. Mthembu (Ed.), *Ethical Research Approaches to Indigenous Knowledge Education* (pp. 83–109). IGI Global. doi:10.4018/978-1-7998-1249-4.ch004

Külcü, Ö. (2020). Verification of Information and Evaluation of Approaches of Information Professionals in Accessing Accurate Information. In K. Dalkir & R. Katz (Eds.), *Navigating Fake News, Alternative Facts, and Misinformation in a Post-Truth World* (pp. 162–183). IGI Global. doi:10.4018/978-1-7998-2543-2.ch007

Kwakye, M. M. (2020). Conceptual Model and Design of Semantic Trajectory Data Warehouse. *International Journal of Data Warehousing and Mining, 16*(3), 108–131. doi:10.4018/IJDWM.2020070106

Lakehal, A., Alti, A., Laborie, S., & Roose, P. (2020). A Semantic Agile Approach for Reconfigurable Distributed Applications in Pervasive Environments. *International Journal of Ambient Computing and Intelligence, 11*(2), 48–67. doi:10.4018/IJACI.2020040103

Lança, M., & Ribeiro, F. B. (2020). Tourism and Sex Outside the "Sex Tourism Paradigm". In P. Pinto & M. Guerreiro (Eds.), *Handbook of Research on Resident and Tourist Perspectives on Travel Destinations* (pp. 361–388). IGI Global. doi:10.4018/978-1-7998-3156-3.ch017

Lang, C. (2020). Paradigm Surfing Across Disciplines in Scholarship and Education Research. In L. Ling & P. Ling (Eds.), *Emerging Methods and Paradigms in Scholarship and Education Research* (pp. 216–229). IGI Global. doi:10.4018/978-1-7998-1001-8.ch012

Latrous, W. O., & Khadraoui, M. (2020). Cultural Challenges of E-Learning Experiences. *International Journal of E-Services and Mobile Applications, 12*(3), 18–37. doi:10.4018/IJESMA.2020070102

Lawal, V., & Bitso, C. (2020). Autoethnography in Information Science Research. In P. Ngulube (Ed.), *Handbook of Research on Connecting Research Methods for Information Science Research* (pp. 114–138). IGI Global. doi:10.4018/978-1-7998-1471-9.ch007

Lebedev, S. V., & Panteleyev, M. (2020). Ontology-Driven Design and Development of Situation Assessment Software in Cyber-Physical Systems. In S. Balandin & E. Balandina (Eds.), *Tools and Technologies for the Development of Cyber-Physical Systems* (pp. 51–76). IGI Global. doi:10.4018/978-1-7998-1974-5.ch003

Lee, D. C., & Gefen, D. (2020). Promises and Challenges of Medical Patient Healthcare Portals in Underserved Communities. In R. W. McHaney, I. Reychev, J. Azuri, M. E. McHaney, & R. Moshonov (Eds.), *Impacts of Information Technology on Patient Care and Empowerment* (pp. 219–251). IGI Global. doi:10.4018/978-1-7998-0047-7.ch012

Lee, W. W. (2020). Ethical Computing for Data Protection. *International Journal of Technoethics, 11*(1), 43–58. doi:10.4018/IJT.2020010104

Lemeni, N. A. (2020). The Relevance of Science-Religion Courses in the Educational Field. In V. Trif (Ed.), *Analyzing Paradigms Used in Education and Educational Psychology* (pp. 306–328). IGI Global. doi:10.4018/978-1-7998-1427-6.ch015

Lemonakis, C., Kouskoukis, M. N., Garefalakis, A., Zopounidis, C., & Eskantar, M. (2020). A Bibliometric Analysis of Ethical Investments (EI) Research. In C. Zopounidis, D. Kenourgios, & G. Dotsis (Eds.), *Recent Advances and Applications in Alternative Investments* (pp. 286–305). IGI Global. doi:10.4018/978-1-7998-2436-7.ch012

Lin, C., Kunnathur, A. S., & Li, L. (2020). The Cultural Foundation of Information Security Behavior. *Journal of Database Management, 31*(2), 21–41. doi:10.4018/JDM.2020040102

Ling, L. (2020). The Power of the Paradigm in Scholarship in Higher Education. In L. Ling & P. Ling (Eds.), *Emerging Methods and Paradigms in Scholarship and Education Research* (pp. 22–55). IGI Global. doi:10.4018/978-1-7998-1001-8.ch002

Livingston, K. (2020). Undertaking Commissioned Research and Scholarship in Education. In L. Ling & P. Ling (Eds.), *Emerging Methods and Paradigms in Scholarship and Education Research* (pp. 199–215). IGI Global. doi:10.4018/978-1-7998-1001-8.ch011

Lombardi, G. (2020). The Role of Unplugged Coding Activity in Developing Computational Thinking in Ages 6-11. In M. Kalogiannakis & S. Papadakis (Eds.), *Handbook of Research on Tools for Teaching Computational Thinking in P-12 Education* (pp. 184–199). IGI Global. doi:10.4018/978-1-7998-4576-8.ch008

Loose, C. (2020). *Teacher Perceptions of Japanese Lesson Study*. IGI Global. doi:10.4018/978-1-7998-4622-2.ch005

Loy, J. (2020). Manufacturing Education for Society 5.0. In S. B. Buckley (Ed.), *Promoting Inclusive Growth in the Fourth Industrial Revolution* (pp. 74–97). IGI Global. doi:10.4018/978-1-7998-4882-0.ch003

Madhani, P. M. (2020). Enhancing Supply Chain Efficiency and Effectiveness With Lean Six Sigma Approach. *International Journal of Project Management and Productivity Assessment*, *8*(1), 40–65. doi:10.4018/IJPMPA.2020010103

Maietti, F., Di Giulio, R., Medici, M., Ferrari, F., Ziri, A. E., Turillazzi, B., & Bonsma, P. (2020). Documentation, Processing, and Representation of Architectural Heritage Through 3D Semantic Modelling. In C. M. Bolognesi & C. Santagati (Eds.), *Impact of Industry 4.0 on Architecture and Cultural Heritage* (pp. 202–238). IGI Global. doi:10.4018/978-1-7998-1234-0.ch009

Makoza, F. (2020). Analysis of Discourses on WhatsApp Coup Reported in the Media. *International Journal of Social Media and Online Communities*, *12*(2), 42–60. doi:10.4018/IJSMOC.2020070103

Makwerere, D., Manyika, R. S., Mutinhima, M., & Saratiere, A. K. (2020). Non-Governmental Organisations and Culturally-Sensitive Gender Programming. In J. Kurebwa (Ed.), *Understanding Gender in the African Context* (pp. 87–104). IGI Global. doi:10.4018/978-1-7998-2815-0.ch005

Malikov, A. (2020). Sukuk as an Infrastructure-Financing Tool. In A. Rafay (Ed.), *Handbook of Research on Theory and Practice of Global Islamic Finance* (pp. 770–787). IGI Global. doi:10.4018/978-1-7998-0218-1.ch039

Malinga, T., & Modie-Moroka, T. (2020). Researching African Diasporas. In U. T. Onyebadi (Ed.), *Multidisciplinary Issues Surrounding African Diasporas* (pp. 1–26). IGI Global. doi:10.4018/978-1-5225-5079-2.ch001

Maluleka, J. R. (2020). Understanding Traditional Healing Practices in the Limpopo Province of South Africa. In P. Ngulube (Ed.), *Handbook of Research on Connecting Research Methods for Information Science Research* (pp. 233–244). IGI Global. doi:10.4018/978-1-7998-1471-9.ch012

Mardi, F. (2020). Gears in Motion. *International Journal of Virtual and Personal Learning Environments, 10*(2), 35–49. doi:10.4018/IJVPLE.2020070103

Marutha, N. S. (2020). Studying Medical Records Management in the Public Healthcare Sector of South Africa Using Multi-Method. In P. Ngulube (Ed.), *Handbook of Research on Connecting Research Methods for Information Science Research* (pp. 404–424). IGI Global. doi:10.4018/978-1-7998-1471-9.ch021

Mavodza, J. (2020). The Evolving Landscape of Research Methods in Library and Information Science. In P. Ngulube (Ed.), *Handbook of Research on Connecting Research Methods for Information Science Research* (pp. 1–17). IGI Global. doi:10.4018/978-1-7998-1471-9.ch001

Mayerhoffer, M. (2020). Post-Hofstedean Approaches to Culture. In R. Brunet-Thornton (Ed.), *Examining Cultural Perspectives in a Globalized World* (pp. 1–22). IGI Global. doi:10.4018/978-1-7998-0214-3.ch001

McDowell, J. (2020). A Theory-Practice Research Framework for Video-Enhanced Learning, Assessment, and Feedback. In C. E. Dann & S. O'Neill (Eds.), *Technology-Enhanced Formative Assessment Practices in Higher Education* (pp. 100–126). IGI Global. doi:10.4018/978-1-7998-0426-0.ch006

Mekonnen, G. A. (2020). Fulfilling the Responsibility to Protect. *International Journal of Risk and Contingency Management, 9*(1), 29–54. doi:10.4018/IJRCM.2020010103

Meriah, I., & Rabai, L. B. (2020). Analysing Information Security Risk Ontologies. *International Journal of Systems and Software Security and Protection, 11*(1), 1–16. doi:10.4018/IJSSSP.2020010101

Miori, V. M., & Herschel, R. T. (2020). Consumer Online Behavior, Data Sharing, and Ethics. In E. C. Idemudia (Ed.), *Handbook of Research on Social and Organizational Dynamics in the Digital Era* (pp. 409–424). IGI Global. doi:10.4018/978-1-5225-8933-4.ch019

Mirarchi, C., Pavan, A., Di Martino, B., & Esposito, A. (2020). Impact of Industry 4.0 in Architecture and Cultural Heritage. In C. M. Bolognesi & C. Santagati (Eds.), *Impact of Industry 4.0 on Architecture and Cultural Heritage* (pp. 306–329). IGI Global. doi:10.4018/978-1-7998-1234-0.ch013

Mittal, S., & Kumar, V. (2020). A Framework for Ethical Mobile Marketing. *International Journal of Technoethics*, *11*(1), 28–42. doi:10.4018/IJT.2020010103

Mladenović, M., Stanković, S. V., & Pajić, V. (2020). Two Ways for the Automatic Generation of Application Ontologies by Using BalkaNet. *International Journal on Semantic Web and Information Systems*, *16*(2), 18–41. doi:10.4018/IJSWIS.2020040102

Moahi, K. H. (2020). The Research Process and Indigenous Epistemologies. In P. Ngulube (Ed.), *Handbook of Research on Connecting Research Methods for Information Science Research* (pp. 245–265). IGI Global. doi:10.4018/978-1-7998-1471-9.ch013

Moise, L. L., Khoase, R., & Ndayizigamiye, P. (2020). The Influence of Government Support Interventions on the Growth of African Foreign-Owned SMMEs in South Africa. In M. Galindo-Martín, M. Mendez-Picazo, & M. Castaño-Martínez (Eds.), *Analyzing the Relationship Between Innovation, Value Creation, and Entrepreneurship* (pp. 104–124). IGI Global. doi:10.4018/978-1-7998-1169-5.ch006

Monticolo, D., & Lahoud, I. (2020). An Agent Approach to Manage Heterogeneous and Distributed Knowledge. *International Journal of Knowledge-Based Organizations*, *10*(1), 27–48. doi:10.4018/IJKBO.2020010102

Moran, R., Roberston, L., Tai, C., Keith, K. J., Price, J., Meier, L. T., & Hong, H. (2020). Preparing Pre-Service Teachers for the Future. In J. Keengwe & P. Wachira (Eds.), *Handbook of Research on Integrating Computer Science and Computational Thinking in K-12 Education* (pp. 17–29). IGI Global. doi:10.4018/978-1-7998-1479-5.ch002

Morrison, J. A., & Kim, J. (2020). Resisting the “Tyranny of an Expert”. In A. S. Zimmerman (Ed.), *Preparing Students for Community-Engaged Scholarship in Higher Education* (pp. 43–60). IGI Global. doi:10.4018/978-1-7998-2208-0.ch003

Mosweu, O., & Mosweu, T. (2020). Research Methodologies Used in Library and Information Studies Masters’ Degree Dissertations at the University of Botswana. In P. Ngulube (Ed.), *Handbook of Research on Connecting Research Methods for Information Science Research* (pp. 384–403). IGI Global. doi:10.4018/978-1-7998-1471-9.ch020

Mountasser, I., Ouhbi, B., Hdioud, F., & Frikh, B. (2020). Ontology-Based Open Tourism Data Integration Framework. In M. Strydom & S. Buckley (Eds.), *AI and Big Data’s Potential for Disruptive Innovation* (pp. 44–70). IGI Global. doi:10.4018/978-1-5225-9687-5.ch002

Moyo, N., & Gonye, J. (2021). Theorising the Politics of Knowledge Production in Curriculum in Zimbabwe. In N. Mthembu (Ed.), *Ethical Research Approaches to Indigenous Knowledge Education* (pp. 178–205). IGI Global. doi:10.4018/978-1-7998-1249-4.ch008

Msallam, M., & Wainakh, M. (2020). Improving Performance of Bell Labs Layered Space-Time Systems by Multiframe Space-Time Interleavers. *International Journal of Embedded and Real-Time Communication Systems, 11*(1), 76–95. doi:10.4018/IJERTCS.2020010105

Munn, C. (2020). Computational Thinking and Robotics. In J. A. Delello & R. R. McWhorter (Eds.), *Disruptive and Emerging Technology Trends Across Education and the Workplace* (pp. 90–115). IGI Global. doi:10.4018/978-1-7998-2914-0.ch004

Musa, S. F., & Idris, D. S. (2020). Addressing Issues of Unemployment in Brunei. *International Journal of Asian Business and Information Management, 11*(2), 88–101. doi:10.4018/IJABIM.2020040106

Musendekwa, M., & Rutoro, E. (2020). Contextualising Theology. In A. D. Queirós (Ed.), *Examining a New Paradigm of Heritage With Philosophy, Economy, and Education* (pp. 71–84). IGI Global. doi:10.4018/978-1-7998-3636-0.ch006

Musyoka, M. M., & Adeoye, S. O. (2021). Designing an Inclusive Culturally Competent Classroom for Immigrant Deaf Students in the United States. In K. Sprott, J. R. O'Connor Jr, & C. Msengi (Eds.), *Designing Culturally Competent Programming for PK-20 Classrooms* (pp. 180–197). IGI Global. doi:10.4018/978-1-7998-3652-0.ch010

Naepi, S. (2020). I Didn't Come to Play. In T. Moeke-Pickering, S. Cote-Meek, & A. Pegoraro (Eds.), *Critical Reflections and Politics on Advancing Women in the Academy* (pp. 52–69). IGI Global. doi:10.4018/978-1-7998-3618-6.ch004

Narayana, A. S., & Bhusal, R. L. (2020). Adoption and Use of Human Information System Digital Technology for Organizational Competitiveness. In E. C. Idemudia (Ed.), *Handbook of Research on Social and Organizational Dynamics in the Digital Era* (pp. 250–275). IGI Global. doi:10.4018/978-1-5225-8933-4.ch012

Narula, S., Kumar, A., Puppala, H., Dwivedy, M., Prakash, S., Singh, R., & Talwar, V. (2020). Restarting Manufacturing Industries Post Covid-19. *International Journal of Strategic Decision Sciences, 11*(2), 46–65. doi:10.4018/IJSDS.2020040103

Naz, G. A. (2020). A Grounded Theory on Conflict Management in Long-Distance Relationships. *International Journal of Sociotechnology and Knowledge Development, 12*(3), 41–59. doi:10.4018/IJSKD.2020070103

Nehinbe, J. O. (2021). Statistical Methods for Conducting the Ontology and Classifications of Fake News on Social Media. In M. Cruz-Cunha & N. R. Mateus-Coelho (Eds.), *Handbook of Research on Cyber Crime and Information Privacy* (pp. 632–651). IGI Global. doi:10.4018/978-1-7998-5728-0.ch029

Ng, V. (2020). Re-Designing the Architecture Curriculum Through the Lens of Graduate Capabilities. In P. Kumar, M. J. Keppell, & C. L. Lim (Eds.), *Preparing 21st Century Teachers for Teach Less, Learn More (TLLM) Pedagogies* (pp. 221–242). IGI Global. doi:10.4018/978-1-7998-1435-1.ch013

Ngulube, B. (2020). The Contribution of Case Study Research in Information Science. In P. Ngulube (Ed.), *Handbook of Research on Connecting Research Methods for Information Science Research* (pp. 95–113). IGI Global. doi:10.4018/978-1-7998-1471-9.ch006

Ngulube, P. (2020). The Movement of Mixed Methods Research and the Role of Information Science Professionals. In P. Ngulube (Ed.), *Handbook of Research on Connecting Research Methods for Information Science Research* (pp. 425–455). IGI Global. doi:10.4018/978-1-7998-1471-9.ch022

Ngulube, P. (2020). Theory and Theorising in Information Science Scholarship. In P. Ngulube (Ed.), *Handbook of Research on Connecting Research Methods for Information Science Research* (pp. 18–39). IGI Global. doi:10.4018/978-1-7998-1471-9.ch002

Nguyen, M. D. (2020). A New Decision-Making Model for Manufacturing Line Designs in Vietnamese Manufacturing Plants. *International Journal of Information Technology Project Management*, *11*(3), 14–29. doi:10.4018/IJITPM.2020070102

Nica, E., Şanta, A. I., Valaskova, K., & Horga, M. G. (2020). Burnout. *International Journal of Sustainable Economies Management*, *9*(3), 21–32. doi:10.4018/IJSEM.2020070102

Niu, Y., & Yao, Y. (2020). Big Data Ethics in Social Media— A Case Study of Evaluating Websites. *International Journal of Organizational and Collective Intelligence*, *10*(1), 14–31. doi:10.4018/IJOCI.2020010102

Nkosinkulu, Z. (2021). A Decolonial Curriculum Is Everything. In N. Mthembu (Ed.), *Ethical Research Approaches to Indigenous Knowledge Education* (pp. 59–81). IGI Global. doi:10.4018/978-1-7998-1249-4.ch003

Novais, R. A. (2020). Organizing With Self-Organization? In A. M. Solo (Ed.), *Handbook of Research on Politics in the Computer Age* (pp. 315–337). IGI Global. doi:10.4018/978-1-7998-0377-5.ch017

Ogata, T. (2020). *Narratology and Post-Narratology*. IGI Global. doi:10.4018/978-1-5225-9693-6.ch003

Ogunmola, G. A., & Kumar, V. (2020). Web Analytics and Online Retail. *International Journal of Technoethics*, *11*(2), 18–33. doi:10.4018/IJT.2020070102

Omitola, T., & Wills, G. (2020). Developing a Framework and an Instrument for Measuring System Openness. *International Journal of Organizational and Collective Intelligence*, *10*(3), 55–96. doi:10.4018/IJOCI.2020070104

Omri, M. N. (2020). Fuzzy Ontology-Based Querying User' Requests Under Uncertain Environment. *International Journal of Cognitive Informatics and Natural Intelligence*, *14*(3), 41–59. doi:10.4018/IJCINI.2020070103

Oncioiu, I., Popovici, D. A., Ungureanu, H. C., & Bîlcan, F. R. (2020). Commercial Real Estate-Specific Approaches and Ethical Dilemmas in the Relationship Between Stakeholders. In I. Oncioiu (Ed.), *Improving Business Performance Through Innovation in the Digital Economy* (pp. 16–27). IGI Global. doi:10.4018/978-1-7998-1005-6.ch002

Onyancha, O. B. (2020). Informetrics Research Methods Outlined. In P. Ngulube (Ed.), *Handbook of Research on Connecting Research Methods for Information Science Research* (pp. 320–348). IGI Global. doi:10.4018/978-1-7998-1471-9.ch017

Orlova, E. (2020). *Theory of Second and Foreign Language Learning*. IGI Global. doi:10.4018/978-1-7998-2672-9.ch001

Orluwene, G. W., & Ajala, I. (2020). Determining Research Design and Data Analysis. In D. Z. Atibuni (Ed.), *Postgraduate Research Engagement in Low Resource Settings* (pp. 36–57). IGI Global. doi:10.4018/978-1-7998-0264-8.ch003

Ormin, K. (2020). Triangulation Approaches in Accounting Research. In M. L. Baran & J. E. Jones (Eds.), *Applied Social Science Approaches to Mixed Methods Research* (pp. 186–200). IGI Global. doi:10.4018/978-1-7998-1025-4.ch008

Orviska, M., & Hunady, J. (2020). E-Commerce in the EU. In K. Jakobs (Ed.), *Shaping the Future Through Standardization* (pp. 226–250). IGI Global. doi:10.4018/978-1-7998-2181-6.ch009

Ouyang, J., Yin, J., & Sun, Y. (2020). Lightweight and Secure Image Segmentation-Based Consensus Mechanism. *International Journal of Systems and Service-Oriented Engineering*, *10*(1), 18–33. doi:10.4018/IJSSOE.2020010102

Oyerinde, Y., & Bankole, F. (2021). Influence of Constant Returns to Scale and Variable Returns to Scale Data Envelopment Analysis Models in ICT Infrastructure Efficiency Utilization. In T. Iyamu (Ed.), *Empowering Businesses With Collaborative Enterprise Architecture Frameworks* (pp. 158–181). IGI Global. doi:10.4018/978-1-5225-8229-8.ch007

Pacho, T. O. (2020). Global Citizenship Education in the Era of Globalization. In J. Keengwe (Ed.), *Handbook of Research on Diversity and Social Justice in Higher Education* (pp. 274–291). IGI Global. doi:10.4018/978-1-7998-5268-1.ch016

Pal, K. (2020). Framework for Reusable Test Case Generation in Software Systems Testing. In C. Pang (Ed.), *Software Engineering for Agile Application Development* (pp. 212–229). IGI Global. doi:10.4018/978-1-7998-2531-9.ch009

Pal, K. (2020). Ontology-Assisted Enterprise Information Systems Integration in Manufacturing Supply Chain. In P. Sahoo (Ed.), *Handbook of Research on Developments and Trends in Industrial and Materials Engineering* (pp. 96–122). IGI Global. doi:10.4018/978-1-7998-1831-1.ch005

Palmeira, M., & Musso, F. (2020). 3Rs of Sustainability Values for Retailing Customers as Factors of Influence on Consumer Behavior. In F. Musso & E. Druica (Eds.), *Handbook of Research on Retailing Techniques for Optimal Consumer Engagement and Experiences* (pp. 421–444). IGI Global. doi:10.4018/978-1-7998-1412-2.ch019

Panachuenvongsak, P., & Ashaye, O. R. (2020). Identifying Key Important Factors Affecting Consumer Purchase Behaviour On Luxury Brand Through the Use of Instagram. In B. Adekunle, H. H. Alharahsheh, & A. Pius (Eds.), *Trends and Issues in International Planning for Businesses* (pp. 89–109). IGI Global. doi:10.4018/978-1-7998-2547-0.ch005

Pandey, J., & Sohani, S. S. (2020). Engaging Stakeholders of Crowdsourced Work. *Journal of Electronic Commerce in Organizations*, *18*(3), 38–52. doi:10.4018/JECO.2020070103

Pandian, R. (2020). Development of Algorithms for Medical Image Compression. In S. R. Nayak & J. Mishra (Eds.), *Examining Fractal Image Processing and Analysis* (pp. 38–56). IGI Global. doi:10.4018/978-1-7998-0066-8.ch003

Pandolfo, L., Spanu, S., Pulina, L., & Grosso, E. (2020). Understanding and Modeling Visitor Behaviours for Enhancing Personalized Cultural Experiences. *International Journal of Technology and Human Interaction*, *16*(3), 24–38. doi:10.4018/IJTHI.2020070102

Panichi, L. (2020). The Case for Qualitative Research Into Language Learning in Virtual Worlds. In M. Kruk & M. Peterson (Eds.), *New Technological Applications for Foreign and Second Language Learning and Teaching* (pp. 47–68). IGI Global. doi:10.4018/978-1-7998-2591-3.ch003

Panwar, A., & Bhatnagar, V. (2020). Sentiment Analysis of Game Review Using Machine Learning in a Hadoop Ecosystem. In L. Gaur, A. Solanki, V. Jain, & D. Khazanchi (Eds.), *Handbook of Research on Engineering Innovations and Technology Management in Organizations* (pp. 145–165). IGI Global. doi:10.4018/978-1-7998-2772-6.ch008

Papadakis, S. (2020). Apps to Promote Computational Thinking Concepts and Coding Skills in Children of Preschool and Pre-Primary School Age. In S. Papadakis & M. Kalogiannakis (Eds.), *Mobile Learning Applications in Early Childhood Education* (pp. 101–121). IGI Global. doi:10.4018/978-1-7998-1486-3.ch006

Papadakis, S., & Kalogiannakis, M. (2020). Exploring Preservice Teachers' Attitudes About the Usage of Educational Robotics in Preschool Education. In M. Kalogiannakis & S. Papadakis (Eds.), *Handbook of Research on Tools for Teaching Computational Thinking in P-12 Education* (pp. 339–355). IGI Global. doi:10.4018/978-1-7998-4576-8.ch013

Papadakis, S., & Kalogiannakis, M. (2020). Learning Computational Thinking Development in Young Children With Bee-Bot Educational Robotics. In M. Kalogiannakis & S. Papadakis (Eds.), *Handbook of Research on Tools for Teaching Computational Thinking in P-12 Education* (pp. 289–309). IGI Global. doi:10.4018/978-1-7998-4576-8.ch011

Papazissi, T., & Chatzinikolaou, F. (2021). Medical and Nursing Civil Liability and Ethics in the Provision of Health Services. In K. Dinas, S. Petousis, M. Kalder, & G. Mavromatidis (Eds.), *Handbook of Research on Oncological and Endoscopical Dilemmas in Modern Gynecological Clinical Practice* (pp. 365–374). IGI Global. doi:10.4018/978-1-7998-4213-2.ch025

Parsons, A. W., & Hathaway, J. I. (2020). Video as a Tool for Guided Reflection in Literacy Specialist Preparation. In R. Karchmer-Klein & K. E. Pytash (Eds.), *Effective Practices in Online Teacher Preparation for Literacy Educators* (pp. 85–106). IGI Global. doi:10.4018/978-1-7998-0206-8.ch005

Pasipamire, N. (2020). Integration in Mixed Methods Research Designs by Graduate Students at the University of Science and Technology. In P. Ngulube (Ed.), *Handbook of Research on Connecting Research Methods for Information Science Research* (pp. 456–478). IGI Global. doi:10.4018/978-1-7998-1471-9.ch023

Petlach, M. (2020). From Philosophical Queries to Greater Dilemmas of Neuroethics. In D. Atli (Ed.), *Analyzing the Strategic Role of Neuromarketing and Consumer Neuroscience* (pp. 25–46). IGI Global. doi:10.4018/978-1-7998-3126-6.ch002

Pinto-Llorente, A. M., & Sánchez-Gómez, M. C. (2020). Perceptions and Attitudes of Future Primary Education Teachers on Technology and Inclusive Education. *Journal of Information Technology Research, 13*(3), 37–57. doi:10.4018/JITR.2020070103

Ponomarev, A., & Shilov, N. (2020). Human-Computer Cloud and Its Applications in E-Tourism. In S. Balandin & E. Balandina (Eds.), *Tools and Technologies for the Development of Cyber-Physical Systems* (pp. 202–235). IGI Global. doi:10.4018/978-1-7998-1974-5.ch008

Posavec, K. (2020). *Methodology and Research*. IGI Global. doi:10.4018/978-1-7998-3680-3.ch004

Poulakis, E., & Politis, P. (2020). Teaching Computational Thinking Unplugged. In M. Kalogiannakis & S. Papadakis (Eds.), *Handbook of Research on Tools for Teaching Computational Thinking in P-12 Education* (pp. 200–236). IGI Global. doi:10.4018/978-1-7998-4576-8.ch009

Priyadarshinee, P. (2020). Examining Critical Success Factors of Cloud Computing Adoption. *International Journal of Decision Support System Technology, 12*(2), 80–96. doi:10.4018/IJDSST.2020040105

Qattan, J., & Al Khasawneh, M. (2020). The Psychological Motivations of Online Conspicuous Consumption. *International Journal of E-Business Research, 16*(2), 1–16. doi:10.4018/IJEBR.2020040101

Queirós, A. D. (2020). How the New Paradigm of Environmental Philosophy and Heritage Create a Tourism Scientific Corpus. In A. D. Queirós (Ed.), *Examining a New Paradigm of Heritage With Philosophy, Economy, and Education* (pp. 20–47). IGI Global. doi:10.4018/978-1-7998-3636-0.ch003

Quinn, R. D., Arnold, A., & Littlewood, K. A. (2020). The Transformative Nature of Community Engagement in the Arts. In A. S. Zimmerman (Ed.), *Preparing Students for Community-Engaged Scholarship in Higher Education* (pp. 379–402). IGI Global. doi:10.4018/978-1-7998-2208-0.ch019

Rahim, M. M., Jabberzadeh, M., & Ilhan, N. (2020). Unit of Analysis in Digitally-Enabled Electronic Procurement Research. In K. Sandhu (Ed.), *Digital Innovations for Customer Engagement, Management, and Organizational Improvement* (pp. 83–103). IGI Global. doi:10.4018/978-1-7998-5171-4.ch005

Rajendran, S., & Prabhu, J. (2020). Learnability of Interestingness with Semantic Similarity and Reasoning in the Knowledge Base of Decision Support Systems. *International Journal of Web Portals*, *12*(1), 26–44. doi:10.4018/IJWP.2020010103

Ramkissoon-Babwah, N., & Racha, S. D. (2020). Social Media Marketing in Trinidad and Tobago. *Journal of Media Management and Entrepreneurship*, *2*(2), 72–83. doi:10.4018/JMME.2020070105

Rascão, J. P. (2020). Freedom of Expression, Privacy, and Ethical and Social Responsibility in Democracy in the Digital Age. *International Journal of Business Strategy and Automation*, *1*(3), 1–23. doi:10.4018/IJBSA.2020070101

Ravyse, W. S., Blignaut, A. S., & Botha-Ravyse, C. R. (2020). Codebook Co-Development to Understand Fidelity and Initiate Artificial Intelligence in Serious Games. *International Journal of Game-Based Learning*, *10*(1), 37–53. doi:10.4018/IJGBL.2020010103

Revell, M. D. (2020). Culturally Restorative Instruction. In B. Arslan-Cansever & P. Önder-Erol (Eds.), *Sociological Perspectives on Educating Children in Contemporary Society* (pp. 229–258). IGI Global. doi:10.4018/978-1-7998-1847-2.ch010

Revell, M. D. (2021). Sustaining Culturally Responsive Teaching Practices. In K. Sprott, J. R. O'Connor Jr, & C. Msengi (Eds.), *Designing Culturally Competent Programming for PK-20 Classrooms* (pp. 218–239). IGI Global. doi:10.4018/978-1-7998-3652-0.ch012

Rich, P., & Browning, S. F. (2020). Using Dr. Scratch as a Formative Feedback Tool to Assess Computational Thinking. In J. Keengwe & P. Wachira (Eds.), *Handbook of Research on Integrating Computer Science and Computational Thinking in K-12 Education* (pp. 220–242). IGI Global. doi:10.4018/978-1-7998-1479-5.ch012

Rinaldi, A. M., Russo, C., & Madani, K. (2020). A Semantic Matching Strategy for Very Large Knowledge Bases Integration. *International Journal of Information Technology and Web Engineering*, *15*(2), 1–29. doi:10.4018/IJITWE.2020040101

Rosa, F. D., Junior, L. A., Bonacin, R., & Jino, M. (2020). SecAOnto. *International Journal of Web Portals*, *12*(2), 57–80. doi:10.4018/IJWP.2020070104

Royakkers, L., & van Est, R. (2020). The New Digital Wave of Rationalization. *International Journal of Technoethics*, *11*(1), 59–74. doi:10.4018/IJT.2020010105

Rusko, R. (2021). Values or Strategy? In I. Oncioiu, S. Căpuşneanu, D. I. Topor, & D. M. Constantin (Eds.), *Sustainability Reporting, Ethics, and Strategic Management Strategies for Modern Organizations* (pp. 63–77). IGI Global. doi:10.4018/978-1-7998-4637-6.ch004

Sableski, M., Rosemary, C. A., & Kinnucan-Welsch, K. (2020). Use of a Metacognitive Tool to Facilitate Teacher Reflection in an Online Graduate Literacy Course. In R. Karchmer-Klein & K. E. Pytash (Eds.), *Effective Practices in Online Teacher Preparation for Literacy Educators* (pp. 65–83). IGI Global. doi:10.4018/978-1-7998-0206-8.ch004

Sabourin, R., & Ahmed, R. (2020). Empathizing With the Criminally Insane. *International Journal of Cyber Behavior, Psychology and Learning*, *10*(1), 54–73. doi:10.4018/IJCBPL.2020010105

Saighi, A., Laboudi, Z., Roose, P., Laborie, S., & Ghoualmi-Zine, N. (2020). On Using Multiple Disabilities Profiles to Adapt Multimedia Documents. *International Journal of Information Technology and Web Engineering*, *15*(3), 34–60. doi:10.4018/IJITWE.2020070103

Saini, D., & Sengupta, S. S. (2020). Ethics Courses Teaching Linkage to Quality Management Education. In N. Baporikar & M. Sony (Eds.), *Quality Management Principles and Policies in Higher Education* (pp. 232–245). IGI Global. doi:10.4018/978-1-7998-1017-9.ch012

Sajja, P. S., & Akerkar, R. A. (2020). Fuzzy Ontology for Requirements Determination and Documentation During Software Development. In V. Pendyala (Ed.), *Tools and Techniques for Software Development in Large Organizations* (pp. 21–44). IGI Global. doi:10.4018/978-1-7998-1863-2.ch002

Sambo, G. K., & Musundire, A. (2020). The Impact of Enhancing Social Media Marketing Knowledge on Customer Attraction and Engagement for University Organizational Growth and Development from the TRACK Theory. *International Journal of Applied Management Theory and Research, 2*(2), 19–40. doi:10.4018/IJAMTR.2020070102

Sandoval-Lucero, E., White, T. D., & Bonacquisti, J. D. (2020). Mentoring and Supervision for Women of Color Leaders in North American Higher Education. *International Journal of Bias, Identity and Diversities in Education, 5*(1), 18–45. doi:10.4018/IJBIDE.2020010102

Sangwan, N., & Bhatnagar, V. (2020). Comprehensive Contemplation of Probabilistic Aspects in Intelligent Analytics. *International Journal of Service Science, Management, Engineering, and Technology, 11*(1), 116–141. doi:10.4018/IJSSMET.2020010108

Sauti, L. (2020). Social Media and Library Collaboration. In C. T. Chisita (Ed.), *Cooperation and Collaboration Initiatives for Libraries and Related Institutions* (pp. 312–334). IGI Global. doi:10.4018/978-1-7998-0043-9.ch016

Schettino, P. (2020). Where Is Hanuman? In G. Guazzaroni & A. S. Pillai (Eds.), *Virtual and Augmented Reality in Education, Art, and Museums* (pp. 311–323). IGI Global. doi:10.4018/978-1-7998-1796-3.ch015

Şchiopu, A. F. (2020). Tourism and Hospitality's Young Workforce. In P. Nistoreanu (Ed.), *New Trends and Opportunities for Central and Eastern European Tourism* (pp. 44–62). IGI Global. doi:10.4018/978-1-7998-1423-8.ch003

Scott, T. L. (2020). Missing and Murdered Indigenous Women and Girls in Canada. In J. O. Ayodele (Ed.), *Global Perspectives on Victimization Analysis and Prevention* (pp. 24–39). IGI Global. doi:10.4018/978-1-7998-1112-1.ch002

Sebastian, P. (2020). Language Centers in the Age of Ubiquitous Technology and Eclectic Methodology. *International Journal of Computer-Assisted Language Learning and Teaching, 10*(3), 67–78. doi:10.4018/IJCALLT.2020070105

Sehlapelo, M. C. (2020). Using Grounded Theory to Develop a Framework for an Information Society in Southern Africa. In P. Ngulube (Ed.), *Handbook of Research on Connecting Research Methods for Information Science Research* (pp. 159–191). IGI Global. doi:10.4018/978-1-7998-1471-9.ch009

Sekiwu, D. (2020). Vocalizing Qualitative Methodologies in Education Research. In D. Z. Atibuni (Ed.), *Postgraduate Research Engagement in Low Resource Settings* (pp. 58–83). IGI Global. doi:10.4018/978-1-7998-0264-8.ch004

Seljan, S., Škof Erdelja, N., Kučiš, V., Dunđer, I., & Pejić Bach, M. (2021). Quality Assurance in Computer-Assisted Translation in Business Environments. In F. Pinarbasi & M. Taskiran (Eds.), *Natural Language Processing for Global and Local Business* (pp. 247–270). IGI Global. doi:10.4018/978-1-7998-4240-8.ch011

Selkrig, M., & Keamy, R. (2020). Becoming Entangled in SoTL and Research Through Supercomplexity. In L. Ling & P. Ling (Eds.), *Emerging Methods and Paradigms in Scholarship and Education Research* (pp. 244–262). IGI Global. doi:10.4018/978-1-7998-1001-8.ch014

Semwanga, A. R., & Atwine, A. M. (2020). Examining the Adoption of ICTs for Weather Monitoring and Climate Change Adaptation in Uganda. *International Journal of ICT Research in Africa and the Middle East, 9*(1), 67–81. doi:10.4018/IJICTRAME.2020010104

Serna, A., & Gerrikagoitia, J. K. (2021). Discovery of Sustainable Transport Modes Underlying TripAdvisor Reviews With Sentiment Analysis. In F. Pinarbasi & M. Taskiran (Eds.), *Natural Language Processing for Global and Local Business* (pp. 180–199). IGI Global. doi:10.4018/978-1-7998-4240-8.ch008

Serrão, A. V. (2020). Towards a Landscape Ethics. In A. D. Queirós (Ed.), *Examining a New Paradigm of Heritage With Philosophy, Economy, and Education* (pp. 10–19). IGI Global. doi:10.4018/978-1-7998-3636-0.ch002

Shah, T. J., & Banday, M. T. (2020). A Review of Contemporary Image Compression Techniques and Standards. In S. R. Nayak & J. Mishra (Eds.), *Examining Fractal Image Processing and Analysis* (pp. 121–157). IGI Global. doi:10.4018/978-1-7998-0066-8.ch006

Shah, T. J., & Banday, M. T. (2020). Empirical Performance Analysis of Wavelet Transform Coding-Based Image Compression Techniques. In S. R. Nayak & J. Mishra (Eds.), *Examining Fractal Image Processing and Analysis* (pp. 57–99). IGI Global. doi:10.4018/978-1-7998-0066-8.ch004

Shange, X. M. (2021). The Ambit of Ethics in the South African Academic Institutions. In N. Mthembu (Ed.), *Ethical Research Approaches to Indigenous Knowledge Education* (pp. 24–58). IGI Global. doi:10.4018/978-1-7998-1249-4.ch002

Sharma, U., Sood, M., Puthooran, E., & Kumar, Y. (2020). A Block-Based Arithmetic Entropy Encoding Scheme for Medical Images. *International Journal of Healthcare Information Systems and Informatics*, *15*(3), 65–81. doi:10.4018/IJHISI.2020070104

Shukla, S., Mohanty, B., & Kumar, A. (2020). Strategizing SCM-M Interface Using DeLone and McLean Model of IS Success and Fuzzy Cognitive Maps. *Journal of Electronic Commerce in Organizations*, *18*(1), 59–88. doi:10.4018/JECO.2020010104

Siau, K., & Wang, W. (2020). Artificial Intelligence (AI) Ethics. *Journal of Database Management*, *31*(2), 74–87. doi:10.4018/JDM.2020040105

Singh, S., & Aswal, M. S. (2020). A State of the Art on Big Data With Semantic Web Technologies. In L. Gaur, A. Solanki, V. Jain, & D. Khazanchi (Eds.), *Handbook of Research on Engineering Innovations and Technology Management in Organizations* (pp. 195–216). IGI Global. doi:10.4018/978-1-7998-2772-6.ch011

Slover, E. M. (2020). A Case for Objectivist Ethics in the 21st Century. *International Journal of Responsible Leadership and Ethical Decision-Making*, *2*(1), 19–31. doi:10.4018/IJRLEDM.2020010102

Smirnov, A., Shilov, N., & Shchekotov, M. (2020). Ontology-Based Modelling of State Machines for Production Robots in Smart Manufacturing Systems. *International Journal of Embedded and Real-Time Communication Systems*, *11*(2), 76–91. doi:10.4018/IJERTCS.2020040105

Solis, M. T., Pearson, J., Dixon, D. P., Blanco, A., & Papp, R. (2020). Terrorism Effects on Businesses Post 9/11. *International Journal of Cyber Warfare & Terrorism*, *10*(1), 15–33. doi:10.4018/IJCWT.2020010102

Stevenson, C. N. (2020). Data Speaks. In M. L. Baran & J. E. Jones (Eds.), *Applied Social Science Approaches to Mixed Methods Research* (pp. 119–144). IGI Global. doi:10.4018/978-1-7998-1025-4.ch006

Stonebraker, P. W., & Hidding, G. J. (2020). Comparing Traditional and Online Instructional Delivery. *International Journal of Online Pedagogy and Course Design, 10*(3), 78–90. doi:10.4018/IJOPCD.2020070105

Su, X., Harrison, N., & Moloney, R. (2020). Students of the Imaginary. *International Journal of Bias, Identity and Diversities in Education, 5*(2), 36–53. doi:10.4018/IJBIDE.2020070103

Sule, D. (2021). Forensic Readiness for Enhanced eDiscovery. In M. Cruz-Cunha & N. R. Mateus-Coelho (Eds.), *Handbook of Research on Cyber Crime and Information Privacy* (pp. 236–255). IGI Global. doi:10.4018/978-1-7998-5728-0.ch013

Sun, Y., Okojie, M. C., Yu, W. W., & Boulder, T. C. (2020). Elementary Students as Digital Makers. In S. P. Huffman, S. Loyless, S. Albritton, & C. Green (Eds.), *Leveraging Technology to Improve School Safety and Student Wellbeing* (pp. 262–280). IGI Global. doi:10.4018/978-1-7998-1766-6.ch015

Susanto, H., Yie, L. F., Setiana, D., Asih, Y., Yoganingrum, A., Riyanto, S., & Saputra, F. A. (2021). Digital Ecosystem Security Issues for Organizations and Governments. In Z. Mahmood (Ed.), *Web 2.0 and Cloud Technologies for Implementing Connected Government* (pp. 204–228). IGI Global. doi:10.4018/978-1-7998-4570-6.ch010

Sweidan, S., El-Bakry, H., & Sabbeh, S. F. (2020). Construction of Liver Fibrosis Diagnosis Ontology From Fuzzy Extended ER Modeling. *International Journal of Decision Support System Technology, 12*(1), 46–69. doi:10.4018/IJDSST.2020010103

Tabaklar, T., & Yildirim, C. (2020). The Development of Servitization Concept in the Era of Industry 4.0 Through SCM Perspective. In U. Hacioglu (Ed.), *Handbook of Research on Strategic Fit and Design in Business Ecosystems* (pp. 593–615). IGI Global. doi:10.4018/978-1-7998-1125-1.ch026

Takhom, A., Usanavasin, S., Supnithi, T., Ikeda, M., Hoppe, H. U., & Boonkwan, P. (2020). Discovering Cross-Disciplinary Concepts in Multidisciplinary Context Through Collaborative Framework. *International Journal of Knowledge and Systems Science, 11*(2), 1–19. doi:10.4018/IJKSS.2020040101

Tanniru, M. R., Xi, Y., & Sandhu, K. (2020). Leadership to Advance Innovation for Digital Healthcare Transformation. In K. Sandhu (Ed.), *Leadership, Management, and Adoption Techniques for Digital Service Innovation* (pp. 1–24). IGI Global. doi:10.4018/978-1-7998-2799-3.ch001

Tavares, M. D., & Dias, A. P. (2020). Research Methods and Methodologies Used in Studies on Social Accounting. In G. Azevedo, J. Oliveira, R. Marques, & A. Ferreira (Eds.), *Tools, Strategies, and Practices for Modern and Accountable Public Sector Management* (pp. 123–150). IGI Global. doi:10.4018/978-1-7998-1385-9.ch005

Taylor, P. (2020). *Coursework in Scholarly Habits of Research Methodology and Dissemination*. IGI Global. doi:10.4018/978-1-7998-2656-9.ch005

Tekinarslan, E., Gürer, M. D., & Akayoğlu, S. (2020). Web-Based Data Collection for Educational Research. In E. Kennedy & Y. Qian (Eds.), *Advancing Educational Research With Emerging Technology* (pp. 152–172). IGI Global. doi:10.4018/978-1-7998-1173-2.ch008

Terrazas-Arellanes, F. E., Strycker, L. A., Walden, E. D., & Knox, C. (2020). Development of a Middle School Online Science Curriculum. In L. A. Tomei & D. D. Carbonara (Eds.), *Handbook of Research on Diverse Teaching Strategies for the Technology-Rich Classroom* (pp. 186–224). IGI Global. doi:10.4018/978-1-7998-0238-9.ch014

Thakur, A. (2020). SORT. *International Journal of Wireless Networks and Broadband Technologies*, *9*(2), 115–142. doi:10.4018/IJWNBT.20200701.oa2 PMID:31863412

Thinyane, M. (2020). Standardizing Social Justice in Digital Health. *International Journal of Standardization Research*, *18*(1), 24–43. doi:10.4018/IJSR.20200101.oa2

Thongmak, M. (2020). The Antecedents of Search Performance. *International Journal of Knowledge Management*, *16*(1), 21–42. doi:10.4018/IJKM.2020010102

Toda, K., Tamada, H., Nakamura, M., & Matsumoto, K. (2020). Capturing Spotaneous Software Evolution in a Social Coding Platform With Project-as-a-City Concept. *International Journal of Software Innovation*, *8*(3), 35–50. doi:10.4018/IJSI.2020070103

Tomaselli, G., Garg, L., Gupta, V., Xuereb, P. A., & Buttigieg, S. C. (2020). Corporate Social Responsibility Application in the Healthcare Sector. *International Journal of Information Systems and Social Change, 11*(1), 11–23. doi:10.4018/IJISSC.2020010102

Topor, F. S. (2020). Managerial Ethics and the Function of Culture in Mexico and the United States. In F. S. Topor (Ed.), *Examining Ethics and Intercultural Interactions in International Relations* (pp. 53–82). IGI Global. doi:10.4018/978-1-7998-2377-3.ch003

Torrisi-Steele, G. (2020). Design Thinking as a Paradigm to Support the Ethical Revival in Higher Education. In V. Wang (Ed.), *Handbook of Research on Ethical Challenges in Higher Education Leadership and Administration* (pp. 156–171). IGI Global. doi:10.4018/978-1-7998-4141-8.ch009

Torrisi-Steele, G. (2020). Ethics in Higher Education Leadership. In V. Wang (Ed.), *Handbook of Research on Ethical Challenges in Higher Education Leadership and Administration* (pp. 1–17). IGI Global. doi:10.4018/978-1-7998-4141-8.ch001

Trad, A. (2020). Applied Mathematical Model for Business Transformation Projects. In E. C. Idemudia (Ed.), *Handbook of Research on IT Applications for Strategic Competitive Advantage and Decision Making* (pp. 269–308). IGI Global. doi:10.4018/978-1-7998-3351-2.ch015

Tran, Y. (2020). Enhancing Preservice Teachers' Confidence and Efficacy in Computer Science. In J. Keengwe & P. Wachira (Eds.), *Handbook of Research on Integrating Computer Science and Computational Thinking in K-12 Education* (pp. 145–163). IGI Global. doi:10.4018/978-1-7998-1479-5.ch009

Travica, B. (2020). *Infostructure, Technology, and Organization*. IGI Global. doi:10.4018/978-1-7998-2760-3.ch004

Trif, V. (2020). Culture of Paradigms in Education and in Educational Psychology. In V. Trif (Ed.), *Analyzing Paradigms Used in Education and Educational Psychology* (pp. 1–21). IGI Global. doi:10.4018/978-1-7998-1427-6.ch001

Tubulingane, B. S. (2020). Competitive Advantage and Student Recruitment at a Namibian University. *International Journal of Technology-Enabled Student Support Services, 10*(2), 1–19. doi:10.4018/IJTESSS.2020070101

Urick, M. J. (2020). The Millennial Label. In B. Christiansen (Ed.), *Global Applications of Multigenerational Management and Leadership in the Transcultural Era* (pp. 263–283). IGI Global. doi:10.4018/978-1-5225-9906-7.ch009

Ursyn, A. (2020). *Bringing Learning and Teaching up to Date*. IGI Global. doi:10.4018/978-1-7998-1651-5.ch005

Valderrama, Y. Z. (2020). Journalists' Safety and Multifaceted Censorship in Colombia. In S. Jamil (Ed.), *Handbook of Research on Combating Threats to Media Freedom and Journalist Safety* (pp. 298–318). IGI Global. doi:10.4018/978-1-7998-1298-2.ch016

Van Harpen, G. (2020). Using Data, Description, and Graphics to Enrich Your Mixed Methods Study. In M. L. Baran & J. E. Jones (Eds.), *Applied Social Science Approaches to Mixed Methods Research* (pp. 107–118). IGI Global. doi:10.4018/978-1-7998-1025-4.ch005

Vargas-Hernández, J. G. (2020). Professional Integrity for Educational Quality in Management Sciences. In N. Baporikar & M. Sony (Eds.), *Quality Management Principles and Policies in Higher Education* (pp. 209–231). IGI Global. doi:10.4018/978-1-7998-1017-9.ch011

Vargas-Iglesias, J. J. (2020). An Epistemology of the Event for the Digital Media. In V. Hernández-Santaolalla & M. Barrientos-Bueno (Eds.), *Handbook of Research on Transmedia Storytelling, Audience Engagement, and Business Strategies* (pp. 115–128). IGI Global. doi:10.4018/978-1-7998-3119-8.ch009

Vasiliev, A. N., Goryachkina, V., & Budnikov, D. (2020). Research Methodology for Microwave-Convective Processing of Grain. *International Journal of Energy Optimization and Engineering*, *9*(2), 1–11. doi:10.4018/IJEOE.2020040101

Vasilyev, A. N., Vasilyev, A. A., Budnikov, D., & Belov, A. (2020). To Methodology Research of Microwave. In V. Kharchenko & P. Vasant (Eds.), *Handbook of Research on Smart Computing for Renewable Energy and Agro-Engineering* (pp. 163–195). IGI Global. doi:10.4018/978-1-7998-1216-6.ch007

Vega-Muñoz, A., & Estrada-Muñoz, C. (2020). Evaluating Technostress to Improve Teaching Performance. In A. Realyvásquez-Vargas, K. C. Arredondo-Soto, G. Hernández-Escobedo, & J. González-Reséndiz (Eds.), *Evaluating Mental Workload for Improved Workplace Performance* (pp. 161–183). IGI Global. doi:10.4018/978-1-7998-1052-0.ch008

Vincent, O., Ikpesu, F., & Dakare, O. (2020). Ethics and CSR Practices for Enduring Corporate Governance Culture. In O. S. Agyemang, A. Ansong, & B. K. Agyei-Mensah (Eds.), *Corporate Governance Models and Applications in Developing Economies* (pp. 205–236). IGI Global. doi:10.4018/978-1-5225-9607-3.ch011

Volk, M., Staegemann, D., Jamous, N., Pohl, M., & Turowski, K. (2020). Providing Clarity on Big Data Technologies. *International Journal of Intelligent Information Technologies*, *16*(2), 49–73. doi:10.4018/IJIIT.2020040103

Vundi, S. K., Chepsoi, A. C., & Matanga, F. K. (2020). An Examination of the Challenges the National Police Service (NPS) Face in Counter Terrorism (CT) Within Nairobi City County, Kenya. *International Journal of Disaster Response and Emergency Management*, *3*(2), 24–34. doi:10.4018/IJDREM.2020070102

Wachira, P., & Liu, X. (2020). Computer Science in Mathematics Preservice Teacher Education. In J. Keengwe & P. Wachira (Eds.), *Handbook of Research on Integrating Computer Science and Computational Thinking in K-12 Education* (pp. 269–281). IGI Global. doi:10.4018/978-1-7998-1479-5.ch014

Wang, H., & Ren, J. (2020). A Semantic Map for Indoor Robot Navigation Based on Predicate Logic. *International Journal of Knowledge and Systems Science*, *11*(1), 1–21. doi:10.4018/IJKSS.2020010101

Wang, V., & Johnson, N. (2020). Work Ethic, Leadership Influence, and Higher Education. In V. Wang (Ed.), *Handbook of Research on Ethical Challenges in Higher Education Leadership and Administration* (pp. 112–130). IGI Global. doi:10.4018/978-1-7998-4141-8.ch007

Weber-Lewerenz, B. (2020). Corporate Digital Responsibility in Construction Engineering. *International Journal of Responsible Leadership and Ethical Decision-Making*, *2*(1), 32–49. doi:10.4018/IJRLEDM.2020010103

Whitney, J., Hultgren, M., Jennex, M. E., Elkins, A., & Frost, E. (2020). Using Knowledge Management and Machine Learning to Identify Victims of Human Sex Trafficking. In M. E. Jennex (Ed.), *Knowledge Management, Innovation, and Entrepreneurship in a Changing World* (pp. 360–389). IGI Global. doi:10.4018/978-1-7998-2355-1.ch014

Wickramasinghe, N. (2020). Trying to Predict in Real Time the Risk of Unplanned Hospital Readmissions. In N. Wickramasinghe (Ed.), *Handbook of Research on Optimizing Healthcare Management Techniques* (pp. 299–310). IGI Global. doi:10.4018/978-1-7998-1371-2.ch022

Wickramasinghe, N., & Goldberg, S. (2020). Facilitating Mobile Initiatives in Healthcare and Wellness Management. In N. Wickramasinghe (Ed.), *Handbook of Research on Optimizing Healthcare Management Techniques* (pp. 350–357). IGI Global. doi:10.4018/978-1-7998-1371-2.ch026

Wickramasinghe, N., & Goldberg, S. (2020). Supporting a Value-Based Healthcare Paradigm With Digital Health and Wellness Personalised Monitoring Solutions. In K. Sandhu (Ed.), *Opportunities and Challenges in Digital Healthcare Innovation* (pp. 57–71). IGI Global. doi:10.4018/978-1-7998-3274-4.ch004

Wickramasinghe, N., & Schaffer, J. L. (2020). Using Intelligent Tools to Support Clinical Decision Making. In N. Wickramasinghe (Ed.), *Handbook of Research on Optimizing Healthcare Management Techniques* (pp. 311–323). IGI Global. doi:10.4018/978-1-7998-1371-2.ch023

Wu, R., Wang, C. L., & Hao, A. (2020). What Makes a Fan a Fan? In C. L. Wang (Ed.), *Handbook of Research on the Impact of Fandom in Society and Consumerism* (pp. 378–396). IGI Global. doi:10.4018/978-1-7998-1048-3.ch018

Xue, X., & Chen, J. (2020). Semi-Automatic Sensor Ontology Matching Based on Interactive Multi-Objective Evolutionary Algorithm. In S. Cheng & Y. Shi (Eds.), *Handbook of Research on Advancements of Swarm Intelligence Algorithms for Solving Real-World Problems* (pp. 27–42). IGI Global. doi:10.4018/978-1-7998-3222-5.ch002

Yadav, R., & Mahara, T. (2020). Exploring the Role of E-Servicescape Dimensions on Customer Online Shopping. *Journal of Electronic Commerce in Organizations, 18*(3), 53–73. doi:10.4018/JECO.2020070104

Yağcı, K., Akçay, S., Efendi, M., & Mutlu Öztürk, H. (2020). Information Security Awareness in Tourism Enterprises. In A. Akbaba & V. Altıntaş (Eds.), *Industrial and Managerial Solutions for Tourism Enterprises* (pp. 251–267). IGI Global. doi:10.4018/978-1-7998-3030-6.ch015

Yalovitsyna, S. E., Volokhova, V. V., & Korzun, D. G. (2020). Smart Museum. In S. Balandin & E. Balandina (Eds.), *Tools and Technologies for the Development of Cyber-Physical Systems* (pp. 236–255). IGI Global. doi:10.4018/978-1-7998-1974-5.ch009

Yang, F. O., Wu, W. V., & Wu, Y. A. (2020). Using a Game-Based Mobile App to Enhance Vocabulary Acquisition for English Language Learners. *International Journal of Distance Education Technologies, 18*(3), 1–24. doi:10.4018/IJDET.2020070101

Yang, H., & Wu, W. (2020). The Effects of Social Learning and Internet Ethics of College Students Engaging in Cyberbullying Behavior in Taiwan. *International Journal of Social Media and Online Communities, 12*(1), 14–22. doi:10.4018/IJSMOC.2020010102

Yang, Z., Van Ngo, Q., & Nguyen, C. X. (2020). Ethics of Retailers and Consumer Behavior in E-Commerce. *International Journal of Asian Business and Information Management, 11*(1), 107–126. doi:10.4018/IJABIM.2020010107

Yaokumah, W. (2020). Predicting and Explaining Cyber Ethics with Ethical Theories. *International Journal of Cyber Warfare & Terrorism, 10*(2), 46–63. doi:10.4018/IJCWT.2020040103

Yassin, M., & Ali, S. A. (2020). Survival of New Institutional Sociology Theory. *International Journal of Customer Relationship Marketing and Management, 11*(1), 50–63. doi:10.4018/IJCRMM.2020010104

Zamazal, O. (2020). A Survey of Ontology Benchmarks for Semantic Web Ontology Tools. *International Journal on Semantic Web and Information Systems, 16*(1), 47–68. doi:10.4018/IJSWIS.2020010103

Zeba, F., Shaheen, M., & Krishnankutty, R. (2020). Hedonic and Utilitarian Values Behind Engagement of Online Consumers. *Journal of Electronic Commerce in Organizations, 18*(3), 1–20. doi:10.4018/JECO.2020070101

Zhang, H., & Dunn, R. A. (2020). The Facebook Me. In R. Kalish (Ed.), *Young Adult Sexuality in the Digital Age* (pp. 176–193). IGI Global. doi:10.4018/978-1-7998-3187-7.ch010

Zhang, Y. (2020). Construction and Application of Regional Medical Information Sharing System Based on Big Data. *International Journal of Information System Modeling and Design, 11*(3), 40–61. doi:10.4018/IJISMD.2020070103

Zhao, R., Huang, Y., Zhou, Y., Yang, M., & Liu, X. (2020). Public Risk Perception Towards Power Generation by Municipal Waste Incineration. In S. Patti & G. Trizzino (Eds.), *Advanced Integrated Approaches to Environmental Economics and Policy* (pp. 87–104). IGI Global. doi:10.4018/978-1-5225-9562-5.ch005

Zheng, R. Z. (2020). Learning With Immersive Technology. In R. Z. Zheng (Ed.), *Cognitive and Affective Perspectives on Immersive Technology in Education* (pp. 1–21). IGI Global. doi:10.4018/978-1-7998-3250-8.ch001

Zúñiga y Postigo, G. (2020). What Economics Can Learn From Ontology. In I. Akansel (Ed.), *Examining the Relationship Between Economics and Philosophy* (pp. 217–238). IGI Global. doi:10.4018/978-1-7998-1037-7.ch011

About the Author

Silas Memory Madondo is a Mixed Methods Research specialist currently working as a Senior Researcher at CeDRE International Africa Pvt in Zimbabwe. He worked as a university lecturer for more than 10 years teaching research methods & statistics, evaluation research and quantitative techniques courses at undergraduate and graduate levels. The author published his first textbook 'Essentials of Social Science Research' in 2016 and published quite a number of articles related to research methodologies, p-values/NHST and evaluation research.

Index

M

N

O

P

Q

R

S

T

www.ingramcontent.com/pod-product-compliance
Lightning Source LLC
LaVergne TN
LVHW081324300926
849161LV00008B/1869